Writing Skills for Scientific Communication

Writing Skills for Scientific Communication is a comprehensive guide designed for researchers, students, and professionals striving for excellence in academic and innovative endeavours. Covering every stage of the research journey, it provides insights into defining research problems, conducting literature reviews, and crafting impactful research papers. The book explores intellectual property rights, the patenting process, and licensing, offering practical strategies for protecting and commercializing innovations. Enriched with real-world case studies, ethical guidelines, and technical writing techniques, this guide offers a holistic approach to mastering research and IP management.

This book:

- Offers a complete and systematic guide to scientific writing, covering various aspects, such as clarity, structure, organization, and language usage, to help readers effectively communicate their research.
- Presents in-depth coverage of how to write research articles, including insights into different sections like the abstract, introduction, methods, results, discussion, and conclusion.
- Illustrates extensive information on how to create and use graphs, tables, and figures to present research data effectively, and enhance the understanding of readers.
- Showcases guidance on tailoring scientific communication for different audiences, including experts in the field, peers, and policymakers.
- Discusses ethical aspects of scientific communication, including plagiarism, data manipulation, and responsible authorship.

Whether you're a novice or an experienced researcher, this book equips you with the essential tools to thrive in the dynamic landscape of scientific communication. It is written for senior undergraduates and graduate students in diverse fields, including mechanical engineering, industrial and manufacturing engineering, electrical engineering, electrical and electronics engineering, and computer engineering.

Writing Skills for Scientific Communication

Kaushik Kumar
Chikesh Ranjan

CRC Press
Taylor & Francis Group
Boca Raton London New York

CRC Press is an imprint of the
Taylor & Francis Group, an **informa** business

Designed cover image: Shutterstock

First edition published 2026
by CRC Press
2385 NW Executive Center Drive, Suite 320, Boca Raton FL 33431

and by CRC Press
4 Park Square, Milton Park, Abingdon, Oxon, OX14 4RN

CRC Press is an imprint of Taylor & Francis Group, LLC

ISBN: 978-1-032-78005-4 (hbk)
ISBN: 978-1-032-78007-8 (pbk)
ISBN: 978-1-003-48582-7 (ebk)

DOI: 10.1201/9781003485827

Typeset in Sabon
by SPi Technologies India Pvt Ltd (Straive)

Contents

Preface

The authors are pleased to present the book *Writing Skills for Scientific Communication*. The book title was chosen to reflect the core purpose of the book, that is, equipping researchers and academics with the essential tools to effectively convey their scientific findings. Clear and precise communication is vital in science, where the accurate dissemination of knowledge impacts understanding, collaboration, and innovation. The title emphasizes the focus on writing as a foundational skill for presenting research, whether through papers, reports, or proposals. It also highlights the importance of tailoring communication to scientific audiences, ensuring clarity, coherence, and credibility. By specifying "scientific communication," the title captures the specialized nature of the skills being addressed, distinguishing the book from general writing guides. This concise and descriptive title resonates with the needs of researchers aiming to excel in their academic and professional pursuits. In simple terms, the book *Writing Skills for Scientific Communication* helps the reader, who could be a scientist, researcher, or anyone interested in science, to become better at expressing their ideas and research findings through writing.

Imagine you have some amazing discoveries or important research that you want to share with the world. This book acts as your guide, showing you how to write in a way that is easy to understand, clear, and engaging. It teaches you how to avoid using confusing jargon and unclear language so that your message reaches a broader audience. The book also helps you understand the proper structure for scientific papers, reports, or other forms of communication, ensuring that your work flows logically and makes sense to others. It provides tips and examples for writing different sections of research articles, such as the introduction, methods, results, and conclusions.

You'll also learn how to use visuals like graphs and tables effectively to present your data in a visually appealing and informative manner. Plus, the book teaches you how to give credit to other researchers by citing their work properly, which is essential in scientific writing.

If English is not your first language or you struggle with grammar, don't worry! The book covers language and grammar tips to help you improve your writing skills and sound more professional.

By the end of the book, you'll feel more confident in your ability to communicate your scientific work clearly and effectively. Whether you want to publish research articles in journals, present at conferences, or share your findings with the public, this book equips you with the necessary skills to make your scientific communication impactful and accessible to a broader audience.

The entire book is segregated into five chapters describing the various modes, such as journal papers, book chapters, books, IPR, and research project proposals, which are currently the buzz ones and are quite tricky but very essential components for degrees, employment, and promotions for students, researchers, and industrialists and are currently being included in the course curriculum of undergraduate courses at different universities and institutions. Each chapter ends with a summarization of the content and some review questions so that the reader can assess their understanding.

Chapter 1 explores the essentials of defining and addressing a research problem, beginning with its central role in any study. It clarifies the meaning of a research problem as the foundation shaping the research's direction, scope, and objectives. Sources of research problems, including practical issues, theoretical gaps, and societal needs, are discussed, alongside types such as descriptive, exploratory, and causal problems. Criteria for a good research problem—clarity, feasibility, significance, and alignment with expertise—are outlined, along with common selection errors like lack of clarity, irrelevance, and ethical oversights. Strategies to avoid these errors include thorough literature reviews, clarity, feasibility checks, and mentor consultations. The chapter addresses defining the scope and setting SMART objectives to ensure focused research goals. It examines approaches to investigation, such as quantitative, qualitative, and mixed methods, and explains data collection using primary and secondary sources. Steps for effective data collection and analysis are highlighted, emphasizing tools and methodologies to ensure validity. Together, these components create a framework for addressing research problems effectively.

From this point, the book dives into a specific class of writing techniques; hence, Chapter 2 provides information on essential practices for conducting effective literature studies to ensure high-quality research and writing. It emphasizes defining clear objectives, utilizing diverse sources, and applying advanced search techniques to gather relevant information. Researchers are urged to critically evaluate sources for validity and reliability; organize findings through thematic, chronological, or comparative methods; and document references properly to maintain clarity and transparency. Literature analysis is highlighted as a tool for identifying trends, gaps, and context within a field, strengthening arguments, and shaping future studies. The chapter also addresses plagiarism, detailing its types, consequences, and prevention through proper citation and ethical practices like confidentiality and informed consent. Technical writing skills are explored, with guidelines for structuring reports and proposals, including objectives, methodology, and budgets. Finally, the chapter provides insights on presenting findings to review committees and effectively addressing feedback, ensuring research integrity and professionalism.

The next chapter of the book, Chapter 3 "Key Skills for Effective Paper Writing," offers a detailed guide to mastering the essential skills for crafting a well-structured and impactful research paper. It explores the key elements of each section, from creating a clear and concise introduction to conducting an analytical literature review and presenting well-organized methods. The chapter underscores clarity and precision in the results, critical thinking in the discussion, and forward-thinking in the conclusions. Emphasis is placed on proper citation, statistical accuracy, logical flow, and effective use of visuals to enhance communication. By addressing these aspects, the chapter equips researchers to present their findings effectively while upholding academic and professional standards.

Chapter 4, "Intellectual Property Rights and the Patenting Process," provides an in-depth overview of how patents protect inventions and promote innovation. It explains the key criteria for patentability—novelty, inventive step, and industrial applicability—and details the steps in the patenting process, from application filing to examination and granting. The chapter also explores international systems like the Patent Cooperation Treaty (PCT) and regional options that simplify global patent protection. While highlighting the benefits of patents, such as exclusive rights and commercial opportunities, it acknowledges the challenges, including high costs, complexity, and difficulties in navigating legal requirements and enforcing patents across jurisdictions.

The last chapter of the book, that is, Chapter 5, explores the complexities of patent rights and their critical role in intellectual property (IP) management, beginning with their scope and importance in safeguarding innovations. It delves into licensing and technology transfer, outlining types of agreements, benefits, and mechanisms for transferring technology between entities, with a focus on patent licensing as a tool for commercializing innovations. Challenges in licensing, including legal, financial, and logistical issues, are discussed alongside the vital role of Technology Transfer Offices (TTOs) in facilitating IP transactions. The chapter also highlights the value of patent databases for research and strategy development and examines emerging trends in IP, particularly in biological systems and computer software. Case studies, such as Apple vs. Samsung, Novartis vs. India, and Google vs. Oracle, illustrate real-world scenarios across diverse IP domains. By blending theoretical insights with practical examples, the chapter offers a comprehensive overview of patent rights, licensing, technology transfer, and the evolving landscape of IP management.

In conclusion, this book is a culmination of years of exploration, research, and personal growth. It is our hope that the insights and knowledge shared within these pages will inspire, challenge, and guide you in your own journey. Whether you are a seasoned professional or a curious learner, we believe that the ideas presented here will resonate and offer value. As you turn the pages, we invite you to engage with the material, question assumptions, and explore new perspectives. Thank you for allowing us to share this work with you—we hope it serves as a meaningful contribution to your endeavours.

Kaushik Kumar
Chikesh Ranjan

About the authors

Kaushik Kumar holds a Ph.D. in Engineering from Jadavpur University, India, an MBA in Marketing Management from Indira Gandhi National Open University, India and a Bachelor of Technology from Regional Engineering College (Now National Institute of Technology), Warangal, India. For 11 years, he worked in a manufacturing unit of Global repute. He is currently working as a Professor in the Department of Mechanical Engineering, Birla Institute of Technology, Mesra, Ranchi, India. He has 23 Years of Teaching and Research Experience. His research interests include Composites, Optimization, Non-Conventional Machining, CAD / CAM, Rapid Prototyping and Quality Management Systems towards product development for societal and industrial usage and has received 40+ patents for them most of which are donated to Start-ups and Industrial fraternity through Jharkhand Small Industries Association (JSIA). He has published 60+ Books (They are referred as Text Books and Reference Books by 40+ Universities / Institutes in their academic curriculum), 100+ Book Chapters and 200+ Research Papers in peer-reviewed reputed national and international journals. Kaushik has also served as Editor–in–Chief, Series Editor, Guest Editor, Editor, Editorial Board Member and Reviewers for International and National Journals. He has been felicitated with many awards and honours including Distinguished Alumnus Award for Professional Excellence 2023 under Academic and Research from National Institute of Technology, Warangal, India. He has also received Sponsored Research and Consultancy Projects of more than 1 Crore from Govt. of India and abroad. Kaushik has delivered expert lectures as Keynote Speaker at International & National Conferences, Resource Person at Various Workshops, FDP's and Short-Term Courses. He has guided many students of Doctoral, Masters and Under Graduate programmes of his home and other institutions in India and abroad. He also has served as Reviewer and Examiner of Doctoral and Masters dissertation for institutes in India and abroad.

Chikesh Ranjan is currently serving as a Project Engineer for Project SwaYaan – Capacity Building in Drone/UAS at the Department of Mechanical Engineering, NIT Rourkela, India. He holds a PhD in Mechanical Engineering from the Department of Mechanical Engineering at the Birla Institute of Technology, Mesra, Ranchi, Jharkhand, India, where he also completed his Master of Engineering (M.E.) in Mechanical Engineering, specializing in the Design of Mechanical Equipment. Additionally, he earned a Bachelor of Engineering (B.E.) in Mechanical Engineering from the Marathwara Institute of Technology, Aurangabad, Maharashtra, India, graduating with First Class with Distinction.

Chikesh Ranjan has made significant contributions to the field of mechanical engineering through his research, having published 17 journal papers, authored 12 books, and written 16 book chapters. He has been actively involved in organizing 10 workshops and coordinating 13 events, in addition to participating in 31 training programs and attending 34 national and international workshops, seminars, and webinars. His innovative work has been recognized with 13 design patents and 9 foreign patents, underscoring his commitment to advancing mechanical design and technology.

Acknowledgements

The creation of *Writing Skills for Scientific Communication* has been a profound and rewarding journey. This book is the culmination of insights, experiences, and the generous support of numerous individuals and institutions who have contributed in various ways to its development, and we would not have been able to complete it without the guidance, cooperation, and encouragement of many individuals.

First and foremost, we would like to thank God for providing the opportunity to believe in passion and hard work and pursue dreams. In the process of putting this book together, it was realized how true this gift of writing is for anyone. This could never have been done without the power provided by you and, of course, the faith in You, the Almighty.

Our deepest gratitude goes to our families for their unending encouragement and understanding. Your patience during the long hours of writing and your unwavering belief in us provided the emotional support necessary to see this project through to completion. Your words of encouragement and your ability to remind me of the bigger picture have been invaluable, which also helped us to take yet another challenge, which decreases the amount of time we could spend with them. They were our inspiration and motivation, and their constant encouragement and late-night brainstorming sessions were invaluable. The unwavering support and insightful feedback helped to shape the direction of this book, and to say the least, their belief in our vision made all the difference. We would like to thank our parents and grandparents for allowing us to follow our ambitions.

We are deeply indebted to Prof. Apurba Kumar Roy of the Department of Mechanical Engineering, Birla Institute of Technology Mesra, Ranchi, India, whose guidance and encouragement have been a constant source of inspiration. Your expertise in scientific communication and your ability to demystify complex ideas have profoundly influenced the approach taken in this book. Your constructive feedback and unwavering support have been invaluable in shaping the direction and content of this work.

We wish to extend our heartfelt thanks to all of our colleagues and friends in different parts of the world for sharing ideas in shaping our thoughts. Your insights and expertise enriched this book in countless ways. It provided thoughtful critique, helpful suggestions, and the benefit of your experiences in scientific writing. Your willingness to share your knowledge and offer constructive feedback has ensured that this book addresses the practical challenges faced by writers in the scientific community.

We are especially grateful to our students, past and present, whose enthusiasm for learning and eagerness to improve their writing skills have been a driving force behind this project. Your questions, challenges, and successes provided the motivation to create a comprehensive resource that serves the needs of aspiring and experienced scientific communicators alike. The workshops, seminars, and one-on-one sessions with you all have been a source of inspiration and learning for us as well.

A special note of appreciation goes to the Reviewers, the Editorial Board Members, the Project Development Editor and the complete team at CRC Press, whose belief in the importance of this project made its publication possible. We would also like to thank them for their constant and consistent guidance, support, and cooperation at all stages of this project. Their effort in every phase, from inception to execution, cannot be expressed in words. Your keen eye and thoughtful suggestions helped bring clarity and polish to this manuscript, and your meticulous attention to detail, editorial expertise, and unwavering commitment to excellence have elevated the quality of this work. A special thank you for providing the occasional push when it was most needed. Thank you for your patience, guidance, and professionalism throughout the publication process. We deeply appreciate your hard work and dedication, support and cooperation rendered to the project.

We also want to acknowledge the contributions of the administrative and technical staff of our institutions for their support to this endeavour. From managing logistics to providing access to resources, your behind-the-scenes efforts ensured the smooth progression of this project.

We owe a debt of gratitude to the larger scientific and academic community. Your work and dedication to advancing knowledge have provided the foundation upon which this book is built. This book is a small contribution to the ongoing dialogue about effective scientific communication, and it is my hope that it will inspire others to continue refining the ways we share and disseminate knowledge.

Finally, we are grateful to our readers, whose curiosity and passion are the real motivation behind this work. It is our hope that this book inspires, educates, or simply entertains in the way we intended. Our efforts will come to a level of satisfaction if the students, researchers, and professionals concerned with all the fields get benefitted.

Thank you to everyone who has played a part in this journey. Your contributions, big and small, are deeply appreciated and forever cherished.

With sincere gratitude,
Kaushik Kumar
Chikesh Ranjan

Chapter 1

Research problem

1.1 INTRODUCTION

A research problem is the foundation of any study, identifying a gap in knowledge, an unresolved challenge, or an unexplored opportunity within a specific domain. It provides a clear focus for the investigation, guiding the research objectives and methodology. By analyzing the current state of knowledge, highlighting limitations, and pinpointing areas requiring improvement or innovation, the problem establishes the need for inquiry. A well-defined research problem not only frames the purpose of the study but also underscores its significance, illustrating the potential contributions to academic, industrial, or societal advancements.

1.2 UNDERSTANDING THE RESEARCH PROBLEM

1.2.1 Meaning of research problem

The meaning of a research problem refers to a specific issue, difficulty, or knowledge gap that a researcher aims to address through a study [1]. It serves as the foundation of a research project, identifying the need for investigation and guiding the formulation of research questions or hypotheses.

In essence, a research problem:

- Highlights what is not well understood or needs clarification.
- Pinpoints why the problem matters—its significance to theory, practice, or societal needs.
- Provides a focus and direction for the research process.

A research problem is the core understanding gap that drives a researcher to undertake a study. It identifies a specific uncertainty, difficulty, or contradiction in existing knowledge or practice. By addressing this problem, researchers aim to advance knowledge, provide solutions to practical challenges, or clarify complex phenomena.

DOI: 10.1201/9781003485827-1

Key aspects of a research problem

- **Identifying what is not well understood or needs clarification**
  - A research problem highlights areas where knowledge is incomplete, ambiguous, or conflicting. It pinpoints questions or issues that have not been adequately explored or explained in the existing literature or practices.

  Example

  Although drone technology is well-developed, there might be limited understanding of how environmental factors like high-altitude winds affect their operational efficiency.
- **Pinpointing why the problem matters**
  - A good research problem justifies its importance by explaining why addressing it is necessary for theory, practice, or societal improvement. This could be due to its:
    - Theoretical significance: Expanding or refining existing knowledge frameworks.
    - Practical implications: Solving real-world issues or improving processes.
    - Societal value: Benefiting communities or addressing pressing social concerns.

  Example

  Researching the impact of autonomous vehicles on urban traffic systems matters because it can lead to smarter urban planning and safer roads.
- **Providing a focus and direction**
  - A clearly defined research problem narrows the scope of the study, ensuring that the research is manageable and targeted. It guides the formulation of specific research questions or hypotheses, ensuring that the study remains relevant and aligned with its objectives.

  Example

  If the research problem is "Understanding the limitations of battery technology in drones," it provides a clear focus on energy storage systems rather than exploring all aspects of drone technology.

1.2.2 The Role of a research problem in a study

- Foundation of research
  - The research problem serves as the starting point of the research process. Without it, the study lacks a defined purpose or direction.
  - It determines the scope of the study, the methodologies to be used, and the nature of the data to be collected.
- **Formulating research questions and hypotheses**
  - A research problem forms the basis for developing research questions or hypotheses. These help articulate what the researcher aims to explore, describe, or test.

Example
For the problem "Limited adoption of renewable energy in rural areas," the research question could be "What are the key barriers to renewable energy adoption in rural communities?"

- **Demonstrating Significance**
  - A well-defined problem clarifies the value of the study, helping stakeholders understand why the research is worth conducting. This is crucial for securing support, funding, or academic approval.
- **Facilitating Analysis and Interpretation**
  - It provides a framework for analyzing data and interpreting results. The problem determines the parameters of what is considered a solution or contribution.

1.3 IDENTIFYING RESEARCH PROBLEMS

1.3.1 Sources of research problem

The sources of a research problem refer to the various origins or triggers that help identify an issue worthy of investigation [2]. These sources ensure that the research problem is relevant and significant and contributes to existing knowledge or practical applications. Below are the primary sources:

- **Practical experience**
  - Real-life challenges faced in professional, organizational, or everyday situations often highlight problems requiring solutions.

  Example
  Difficulty in optimizing drone flight stability in harsh weather conditions could lead to a research problem in aerodynamics.
- **Literature review**
  - A thorough review of existing research can reveal gaps, contradictions, or unanswered questions in a specific field.

  Example
  A review of articles on renewable energy might show a lack of studies on efficient storage solutions for offshore wind energy.
- **Theoretical frameworks**
  - Issues or inconsistencies within established theories may lead to the formulation of research problems that aim to refine, validate, or extend them.

  Example
  Testing the applicability of classical economic theories in modern digital marketplaces.
- **Technological advancements**
  - Emerging technologies or innovations often bring new opportunities or challenges that warrant investigation.

Example
The rise of artificial intelligence raises ethical and practical questions about its application in healthcare decision-making.

- **Social and societal issues**
  - Problems related to societal needs, changes, or conflicts can provide meaningful research topics.

Example
Investigating the effects of remote learning on students' mental health during global disruptions such as pandemics.

- **Professional literature and industry needs**
  - Challenges documented in industry reports, white papers, or case studies may suggest research-worthy problems.

Example
Reports highlighting delays in supply chain management can prompt research on predictive algorithms for demand forecasting.

- **Emerging trends**
  - Trends in culture, economy, technology, or politics often uncover areas requiring academic exploration.

Example
Studying the impact of cryptocurrency on traditional banking systems due to its growing popularity.

- **Personal interest or curiosity**
  - Personal passions or intellectual curiosity about certain phenomena can inspire research topics.

Example
An interest in environmental conservation could lead to studying the efficacy of biodegradable materials in packaging.

- **Policy and legislation**
  - New laws, policies, or international agreements may raise questions about their impact or effectiveness.

Example
Researching the effects of stricter data privacy regulations on small businesses.

- **Interdisciplinary exploration**
  - Combining insights from different fields may highlight unique problems or perspectives.

Example
Investigating the psychological effects of virtual reality tools in medical training by blending psychology and engineering.

- **Contradictions in observations**
  - Observing discrepancies between theoretical predictions and real-world outcomes can inspire investigation.

Example
Examining why certain sustainable energy initiatives fail in regions with high renewable potential.

- **Discussions and expert consultations**
  - Engaging with mentors, industry professionals, or academics often reveals areas ripe for exploration.

  Example

  Conversations with engineers on the challenges of additive manufacturing could uncover unaddressed research topics.

1.3.2 Types of research problems

Research problems can be classified into three main categories, each serving distinct purposes across disciplines [3]. These include theoretical research problems, applied research problems, and action research problems.

- **Theoretical research problems**

These problems focus on conceptual and intellectual inquiries, aiming to enhance understanding of abstract ideas, phenomena, or frameworks without necessarily relying on empirical data [3]. They often explore foundational questions, relationships between concepts, or the development of new theories.

Subtypes in social sciences

- **Casuist problems:** Relate to ethical dilemmas or moral decision-making.
- **Difference problems:** Compare or contrast two or more phenomena.
- **Descriptive problems:** Aim to describe a situation or state.
- **Relational problems:** Investigate relationships between variables or characteristics.

Examples

- Exploring ethical frameworks to justify artificial intelligence and machine learning decisions, particularly in moral and autonomous contexts.
- Using mathematical models to explain the evolution of complex traits like intricate anatomical structures or behaviours.

- **Applied research problems**

These problems address practical, real-world challenges and aim to develop tangible solutions that improve societal, technological, environmental, or health-related aspects. They focus on immediate applications and outcomes.

Examples

- Investigating precision agriculture techniques to enhance crop yields while reducing resource waste.

- Designing an energy-efficient and sustainable urban transportation system to lower carbon emissions.

- **Action research problems**

Action research problems aim to foster positive change within specific settings by actively engaging stakeholders. This approach involves implementing interventions, assessing outcomes, and iteratively improving solutions. It is highly collaborative and context-specific.

Examples

- Partnering with healthcare providers to identify barriers to patient medication adherence and implementing strategies to improve compliance.
- Working with non-profit organizations to evaluate and enhance job training programmes for underserved populations.

1.4 CHARACTERISTICS OF A GOOD RESEARCH PROBLEM

1.4.1 Criteria of a good research problem

A good research problem serves as the foundation of meaningful and impactful research [4]. It should meet specific criteria to ensure clarity, relevance, and feasibility. Below are the essential characteristics of a good research problem:

- **Clarity and specificity**
  - The problem must be clearly defined, concise, and unambiguous [5].
  - It should outline what the study intends to address without room for misinterpretation.

  Example
  Instead of "Studying pollution," a clearer problem is "Analyzing the effects of plastic waste on marine biodiversity in the Pacific Ocean."
- **Relevance**
  - The problem should be significant and address a real gap or need in knowledge, practice, or society [6].
  - It must align with the interests of stakeholders (e.g., academics, practitioners, or policymakers).

  Example
  Investigating renewable energy storage solutions is relevant due to the global push toward sustainable energy systems.

- **Researchability**
  - The problem must be researchable using available methods, tools, and resources.
  - It should allow for systematic investigation and data collection within the constraints of time, budget, and expertise.

  Example

  "Assessing the impact of social media on youth mental health" is researchable if surveys or behavioural studies are possible.
- **Novelty and originality**
  - A good research problem should address an unexplored area, fill a gap in the literature, or challenge existing assumptions.
  - However, it should avoid redundancy unless replicating studies for validation or updates.

  Example

  Exploring the ethical implications of AI in decision-making is novel in light of recent advancements.
- **Feasibility**
  - The problem must be feasible in terms of scope and scale, given the resources, expertise, and time available to the researcher.
  - Avoid problems that are too broad or too complex to tackle effectively.

  Example

  Studying "Global climate change" is too broad; narrowing it to "Effects of climate change on coastal communities in Southeast Asia" is more feasible.
- **Alignment with research goals**
  - The problem should align with the researcher's interests, expertise, and the overall objectives of their field of study [7].
  - It must also fit within the larger framework of the research institution or sponsor's agenda.

  Example

  A mechanical engineer may focus on improving drone propulsion systems as part of broader aerospace innovation.
- **Contribution to knowledge**
  - The problem should aim to make a valuable contribution to existing knowledge or provide actionable solutions.
  - It should answer a "so what?" question—why the problem is worth studying.

  Example

  Understanding how urban heat islands affect local weather patterns contributes to climate adaptation strategies.
- **Ethical considerations**
  - The research problem should respect ethical standards, avoiding harm to participants, the environment, or society.
  - Sensitive topics must be approached with care and proper permission.

Table 1.1 Impact of employee training on productivity

Criteria	Description
Clarity	Well-defined and unambiguous.
Relevance	Significant and addresses real needs or gaps.
Researchability	Can be investigated using available methods and resources.
Novelty	Original or builds on existing knowledge meaningfully.
Feasibility	Realistic in terms of scope, time, and resources.
Alignment	Matches the researcher's expertise and institutional goals.
Contribution	Adds value to knowledge or practice.
Ethics	Respects ethical standards and avoids harm.
Simplicity	Manageable scope; not overly complex.
Measurability	Leads to clear and measurable outcomes.

Example
Studying patient medical records to improve treatment protocols must ensure confidentiality and compliance with ethical guidelines.

- **Simplicity and manageability**
  - The problem should be manageable and focused enough to allow the researcher to explore it thoroughly.
  - Complex problems can be broken into smaller, more manageable sub-problems.

  Example
  Instead of "Understanding poverty," the problem could be narrowed to "Investigating the impact of microfinance on rural poverty in India."
- **Measurable outcomes**
  - A good research problem should lead to outcomes that can be measured, analyzed, or evaluated.

  Example
  "How does employee training affect productivity?" allows for measurable outcomes like productivity metrics before and after training, as shown in Table 1.1.

1.5 COMMON ERRORS IN SELECTING A RESEARCH PROBLEM

Selecting a research problem is a critical step in the research process, and errors in this stage can compromise the quality and relevance of the study [8]. Below are common errors made when selecting a research problem.

1.5.1 Lack of clarity

Lack of clarity occurs when a research problem is vaguely defined or overly broad. It makes it difficult to establish specific objectives, leading to ambiguous research questions and scattered efforts. This often results in inconclusive findings and an unfocused study. To avoid this, the research problem should be clearly articulated with well-defined boundaries and objectives.

Example of lack of clarity

Vague research problem

"How can technology improve education?"

- **Why it lack clarity?**
  - It is too broad and doesn't specify the type of technology, educational level (primary, secondary, or higher education), or the aspect of education (teaching methods, access, assessment, etc.).
  - This makes it hard to focus the research, choose appropriate methodologies, or draw meaningful conclusions.

Clear Research Problem

"How can augmented reality (AR) improve science learning outcomes for high school students?"

- **Why it is clear?**
  - Specifies the technology (AR), the target group (high school students), and the specific aspect of education (science learning outcomes).
  - This allows for a focused approach, measurable objectives, and actionable results.

1.5.2 Overly ambitious scope

Overly ambitious scope occurs when a research problem is too broad or complex to address within the available time, resources, or expertise. For example, attempting to study *"The impact of climate change on all ecosystems globally"* is unrealistic for a single researcher. This can lead to incomplete analysis and superficial findings. A manageable scope, such as *"The impact of climate change on coral reefs in the Pacific Ocean,"* ensures depth and feasibility.

Example of overly ambitious scope

Overly broad research problem

"What are the effects of urbanization on the environment worldwide?"

- **Why it is overly ambitious?**
  - Covers a vast geographical area (worldwide).
  - Includes numerous environmental aspects (air, water, biodiversity, etc.).

  - Requires extensive time, resources, and multidisciplinary expertise.

Refined and feasible research problem

"What are the effects of urbanization on air quality in New Delhi over the past decade?"

- **Why it is feasible?**
  - Focuses on a specific location (New Delhi).
  - Targets one aspect of the environment (air quality).
  - Limits the timeframe to a manageable period (past decade).

1.5.3 Repetition of previous studies

Repetition of Previous Studies happens when a researcher selects a problem that has already been thoroughly studied without offering new perspectives, methods, or insights. For example, reanalyzing "The role of social media in youth behaviour" without addressing new platforms or trends adds little value [4]. To avoid this, identify gaps, propose novel approaches, or focus on unexplored contexts to ensure the research contributes to existing knowledge.

Example of repetition of previous studies

Repetitive research problem

"How does social media influence the academic performance of college students?"

- **Why it is repetitive?**
  - This topic has been extensively studied, with numerous papers exploring the relationship between social media and academic performance in college students.
  - If no new context or perspective is introduced, the research lacks originality.

Novel approach to avoid repetition

"How does the use of short-form video platforms like TikTok impact study habits among high school students during exam preparation?"

- **Why it is original?**
  - Focuses on a specific and newer social media platform (TikTok).
  - Targets a different demographic (high school students) and a unique context (exam preparation).
  - Addresses a specific behaviour (study habits) that may not have been studied extensively.

1.5.4 Irrelevance to field or audience

Irrelevance to Field or Audience occurs when a research problem lacks significance or does not address the interests, needs, or current trends of the intended field or audience. For example, studying *"The dietary habits of medieval monks"* may not be relevant to contemporary nutrition science. To avoid this, ensure the research aligns with the field's priorities and solves real-world problems or contributes meaningfully to ongoing discourse.

Example of irrelevance to field or audience

Irrelevant research problem

"The impact of 18th-century poetry on modern agricultural practices."

- **Why it is irrelevant?**
  - There is no clear connection between 18th-century poetry and agriculture.
  - Neither field (literature or agriculture) would find this topic useful or significant.
  - The study does not address any pressing issue or gap in knowledge.

Relevant research problem

"The role of modern agricultural technologies in improving crop yield under climate change conditions."

- **Why it is relevant?**
  - Addresses a critical issue in agriculture (climate change and crop yield).
  - Aligns with current challenges and priorities in the field.
  - Offers practical implications for researchers, policymakers, and farmers.

1.5.5 Personal bias

Personal bias in selecting a research problem occurs when a researcher chooses a topic based on personal preferences, beliefs, or convenience rather than its academic or societal importance. For example, focusing on *"Why my favorite teaching method is the best"* reflects bias and may lead to skewed results. To avoid this, base the problem on objective criteria such as relevance, significance, and gaps in the literature, ensuring impartiality and credibility.

Example of personal bias

Biased research problem

"Why my preferred smartphone brand is the best for productivity."

- **Why it is biased?**
  - The researcher's personal preference for a particular brand influences the choice of the topic.
  - The study may not objectively evaluate other brands or consider all relevant factors (e.g., price, usability, and features).
  - This limits the study's credibility and generalizability.

Unbiased Research Problem

"How do smartphone features like battery life and screen size affect productivity in remote work environments?"

- **Why it is unbiased?**
  - The topic focuses on measurable, relevant features (battery life and screen size) without favouring any particular brand.
  - It examines factors that impact productivity objectively, allowing for a comprehensive and impartial analysis.

1.5.6 Insufficient background research

Insufficient background research occurs when a researcher fails to thoroughly review existing literature or prior studies related to the topic. This can lead to repeating known findings, overlooking key theories or methods, or missing gaps in knowledge. For example, researching *"The effects of exercise on mental health"* without reviewing existing studies could result in redundant conclusions. Proper background research ensures the problem is novel, builds on previous work, and addresses existing gaps [5].

Example of insufficient background research

Research problem without adequate background

"Does listening to classical music improve concentration during studying?"

- **Why it lacks background research?**
  - The researcher may not have reviewed existing studies on the effects of music on concentration, leading to redundant or unoriginal findings.
  - There could be numerous studies already exploring this relationship, and a lack of background research would miss important theories, methods, and results from previous work.

Research problem with adequate background

"How does listening to classical music affect concentration in students during high-stress exam periods?"

- **Why it is well-researched?**

- The researcher reviews existing studies on music and concentration, identifies gaps (e.g., focusing on high-stress periods), and examines specific conditions (exam stress).
- This builds on existing knowledge and addresses an underexplored area, contributing new insights to the field.

1.5.7 Overdependence on available data

Overdependence on available data occurs when a researcher selects a problem solely because data is easily accessible, rather than its relevance or significance. For example, conducting research on *"The relationship between social media usage and sleep patterns"* just because data from a specific app is readily available may limit the scope of the research and may not address more pressing or relevant issues [6]. To avoid this, ensure the research problem aligns with academic objectives and real-world significance, rather than relying on convenience.

Example of overdependence on available data

Research problem with overdependence

"How does daily screen time affect the mental health of teenagers?" (Using data collected from a popular fitness app that tracks screen time)

- **Why it is dependent on available data?**
  - The researcher selected the topic simply because data is readily available from the fitness app.
  - The app may not capture important variables (e.g., the content of screen time or individual mental health factors), and the research may miss crucial nuances.
  - This limits the depth of the study and could result in a shallow or incomplete understanding.

Research problem with better justification

"How does the content and duration of screen time affect mental health outcomes in teenagers during the pandemic?"

- **Why it is more justified?**
  - The problem is rooted in an important societal issue (mental health and the pandemic) rather than convenience.
  - Data may be collected from multiple sources, including surveys and interviews, offering richer insights.
  - The focus on content and duration provides a deeper understanding of the relationship.

1.5.8 Misalignment with researcher's expertise

Misalignment with the researcher's expertise occurs when a researcher chooses a topic outside their area of knowledge or skill set. This can lead to challenges in data analysis, theoretical understanding, and research execution [7]. For example, a researcher with a background in psychology may struggle to conduct a complex study on advanced statistical methods in physics. To avoid this, researchers should select topics that align with their academic background, training, and resources.

Example of misalignment with researcher's expertise

Research problem misaligned with expertise

"How do quantum computing algorithms optimize machine learning models?"

- **Why it is misaligned?**
  - A researcher with a background in psychology or behavioural science may lack the necessary expertise in computer science, quantum computing, and machine learning to address this problem.
  - They may struggle with complex mathematical models and technical details, impacting the quality and accuracy of the research.

Research Problem Aligned with Expertise

"How do cognitive biases influence decision-making in social contexts?"

- **Why it aligns?**
  - A researcher with a background in psychology or behavioural science is well-equipped to study cognitive biases, as they fall within their area of expertise.
  - They are familiar with research methods, theories, and existing literature in this field, making the research more feasible and impactful.

1.5.9 Ignoring feasibility

Ignoring feasibility occurs when a researcher selects a problem without considering whether it is achievable given the available resources, time, or tools. For example, planning a large-scale, longitudinal study with limited funding or a short timeline may lead to incomplete or unmanageable results. To avoid this, researchers should assess factors like data availability, required expertise, budget, and time constraints before selecting a research problem.

Example of ignoring feasibility

Research problem ignoring feasibility

"Conducting a global survey on the impact of climate change on agriculture in all developing countries over the next year."

- **Why it is not feasible?**
  - The scope is too large (all developing countries) for one researcher or even a small team.
  - Gathering data from such a wide range of countries would require significant time, funding, and resources that may not be available.
  - The timeline (one year) may not allow for proper data collection, analysis, and reporting.

Feasible research problem

"Examining the impact of climate change on crop yields in a specific region, such as sub-Saharan Africa, over the past decade."

- **Why it is feasible?**
  - The scope is more focused (one region), making data collection more manageable.
  - The timeline is realistic for data gathering and analysis.
  - Resources (e.g., data sets and government reports) may be more accessible for a focused regional study.

1.5.10 Overlooking ethical considerations

Overlooking ethical considerations occurs when a researcher fails to account for the ethical implications of their research, such as privacy, consent, and potential harm to participants or communities. For example, conducting a study on sensitive health topics without proper informed consent or safeguarding participants' privacy would be unethical [8]. To avoid this, researchers must ensure their work adheres to ethical standards, including obtaining consent, ensuring confidentiality, and considering the potential impact of their findings on participants.

Example of overlooking ethical considerations

Research problem overlooking ethics

"Conducting a study on the mental health of individuals with substance abuse without obtaining informed consent."

- **Why it overlooks ethics?**
  - Participants are not given the opportunity to fully understand the nature of the study or voluntarily agree to participate, violating their rights.

- The study may also inadvertently cause harm to vulnerable individuals by discussing sensitive topics without appropriate safeguards.

Ethical research problem

"Investigating the mental health of individuals recovering from substance abuse with informed consent, confidentiality, and support resources."

- **Why it adheres to ethics?**
  - Informed consent is obtained, ensuring participants understand the study's purpose and their rights.
  - Ethical considerations, such as providing resources for participants' well-being, are incorporated.
  - Privacy and confidentiality are maintained throughout the study.

1.6 AVOIDING ERRORS IN RESEARCH PROBLEM SELECTION

1.6.1 Conduct thorough literature reviews

Conduct thorough literature reviews to understand the existing research, identify gaps, and avoid redundancy [9]. This involves analyzing relevant studies, theories, and methodologies to build a solid foundation for your research problem. It ensures your study is original and informed and contributes meaningfully to the field.

Example of conducting thorough literature reviews

Without a literature review

A researcher selects the problem: *"How does exercise improve mental health?"*

- **Risk:** This topic is extensively studied, and the researcher may unknowingly replicate existing work without offering new insights.

With a thorough literature review

After reviewing studies, the researcher identifies a gap: *"How does high-intensity interval training (HIIT) affect anxiety levels in young adults compared to traditional cardio?"*

- **Why it is better?**
  - The review reveals HIIT is underexplored for anxiety management.
  - The refined problem focuses on a specific exercise type, demographic, and mental health outcomes, ensuring originality and relevance.

1.6.2 Align the problem with current trends and needs

Aligning the problem with current trends and needs ensures your research is relevant and impactful. Focus on topics addressing pressing issues, advancements, or societal demands in your field. For example, studying *"AI applications in healthcare"* aligns with modern technological and healthcare challenges, increasing the study's significance and applicability.

Example of aligning the problem with current trends and needs

Misaligned research problem

"The role of typewriters in office productivity."

- **Why it is outdated?**: Typewriters are obsolete, and the topic is irrelevant to current technological advancements and workplace practices.

Aligned research problem

"The impact of AI-powered tools on productivity in remote work environments."

- **Why it aligns?**
  - Reflects the growing use of AI in modern workplaces.
  - Addresses a relevant issue in the post-pandemic shift to remote work.
  - Meets the needs of industries and academia by exploring current challenges and opportunities.

1.6.3 Ensure clarity and feasibility

Ensure clarity and feasibility by defining a specific, well-focused research problem that can be realistically addressed within the available time, resources, and expertise. This prevents ambiguity and ensures the study is manageable and achievable, leading to meaningful and actionable results.

Example of ensuring clarity and feasibility

Unclear and unfeasible problem

"What are the effects of technology on education globally?"

- **Why it lacks clarity and feasibility?**
  - Too broad, covering all forms of technology and education worldwide.
  - Requires extensive resources, time, and expertise, making it unrealistic for a single researcher.

Clear and feasible problem

"How do virtual reality tools enhance science learning outcomes for high school students in urban schools?"

- **Why it is clear and feasible?**
  - Specifies the technology (virtual reality), target group (high school students), and setting (urban schools).
  - The scope is manageable with available resources and a defined focus, making the research achievable.

1.6.4 Consult mentors or experts for feedback

Consulting mentors or experts for feedback helps refine your research problem, ensuring it is relevant, clear, and feasible. Their experience and knowledge can identify potential issues, suggest improvements, and provide valuable insights to strengthen your study's focus and methodology.

Example of consulting mentors or experts for feedback

Without consulting a mentor

A student chooses the research problem: *"The impact of 3D printing on all industries."*

- **Issue:** The topic is overly broad and lacks focus, making it difficult to address effectively.

With mentor feedback

After consulting a mentor, the problem is refined to: "The impact of 3D printing on prototyping efficiency in the automotive industry."

- **Why it improved?**
  - The mentor suggested narrowing the scope to a specific industry (automotive) and aspect (prototyping efficiency).
  - The revised problem is more focused, feasible, and aligned with the student's resources and expertise.

1.6.5 Evaluate the problem's significance and potential impact

Evaluating the problem's significance and potential impact involves assessing how the research will contribute to knowledge, address real-world issues, or influence policy, practice, or innovation. A significant research problem should:

1. **Address a gap:** Fill in missing knowledge or unanswered questions in the field.

2. **Solve real problems:** Tackle pressing societal, technological, or environmental challenges.
3. **Advance the field:** Introduce new concepts, methods, or applications.
4. **Have broader implications:** Impact a wide audience or lead to future research opportunities.

Example of evaluating the problem's significance and potential impact:

Research Problem

"How can artificial intelligence improve customer service in retail businesses?"

- **Evaluating significance**
  - **Gap in knowledge:** While AI in customer service has been explored, the focus on specific retail challenges (e.g., handling large customer volumes during sales seasons) might be under-researched.
  - **Real-world problem:** Retail businesses face increasing competition, and AI could enhance customer satisfaction, reduce operational costs, and improve service efficiency.
  - **Broader implications:** Could influence retail industry practices and policy, lead to AI-driven innovations, and shape customer service strategies across sectors.
- **Evaluating potential impact**
  - **Practical solutions:** AI could streamline customer service, improving response times and accuracy, which would have immediate benefits for businesses.
  - **Long-term influence:** This research could inspire further AI applications in other sectors like healthcare or education, leading to broader technological advancements.

1.7 SCOPE OF RESEARCH PROBLEM

The scope of a research problem defines the boundaries of the study. It specifies what the research will cover and what it will not. This includes:

1.7.1 Geographical limits

Geographical limits refer to the specific location or area where a research study is conducted. This can include countries, regions, cities, or even specific sites like schools or companies [10]. Defining geographical limits helps to focus the research, ensuring that the findings are relevant to the chosen location and making the study more manageable and contextually appropriate.

Example of geographical limits

Research problem

"The impact of air pollution on public health in metropolitan cities."

- **Without geographical limits**
  - The research could attempt to analyze air pollution and health effects across the entire world, which would be too broad and difficult to manage.
- **With geographical limits**
  - The research focuses specifically on *"The impact of air pollution on public health in Delhi, India, during the past 5 years."*
  - **Why it is focused?**
    - Delhi has a well-known issue with air pollution, and focusing on this city allows for more precise data collection and analysis.
    - It narrows the scope to a specific geographic area, making the study feasible and contextually relevant to the population of that region.

1.7.2 Time frame

Time frame refers to the specific period during which the research will take place or the time span of the data being studied [11]. It defines the starting and ending points of the research, ensuring that the study remains focused and manageable. The time frame could refer to historical data (e.g., studying trends over the last decade) or a period during which the research is conducted (e.g., a 6-month study of a particular intervention). Defining the time frame helps in planning, data collection, and ensuring the research is relevant to a specific period.

Example of time frame

Research problem

"The impact of online learning on student performance during the COVID-19 pandemic."

- **Without a time frame**
  - The research could be too broad, trying to analyze online learning's impact over an undefined period, which could lead to ambiguity in the findings and scope.
- **With a defined time frame**
  - The research focuses specifically on *"The impact of online learning on student performance in high school students from March 2020 to December 2020."*
  - **Why it is focused?**

- The time frame captures the period when online learning became widespread due to the pandemic, making the study more relevant and manageable.
- The research is time-bound, ensuring that data collection, analysis, and conclusions are tied to this specific period.

1.7.3 Population or sample

Population or sample refers to the group of individuals or units that a research study focuses on [12].

- **Population:** The entire group that the research aims to draw conclusions about (e.g., all high school students in a country).
- **Sample:** A smaller, manageable subset of the population selected for the study (e.g., high school students from 10 schools in a specific city).

Example of population or sample

Research problem

"The effect of exercise on mental health in adults."

- **Population**
  - The entire group the research aims to study could be *"All adults aged 18-65 living in urban areas."*
  - **Why it's broad:** This is the entire population the study seeks to understand, but it may be too large to study comprehensively.
- **Sample**
  - A more manageable group could be *"Adults aged 18-65 from five urban centers in the United States."*
  - **Why it's focused?:** This sample is a smaller subset of the population, selected for practicality and relevance. It's easier to manage and still offers insights into the larger population.

1.7.4 Variables

Variables are the factors or characteristics that a researcher studies and measures in a research project. They can be:

- **Independent variables:** The factors that are manipulated or changed to observe their effect (e.g., exercise frequency).
- **Dependent variables:** The outcomes that are measured to see how they are affected by changes in the independent variables (e.g., mental health status).
- **Control variables:** Other factors that are kept constant to ensure the results are due to the independent variable alone (e.g., age and diet).

Example of variables

Research problem

"The effect of exercise on mental health in adults."

- **Independent variable**
  - *Exercise frequency* (e.g., number of hours per week spent exercising).
  - **Why it's independent?**: The researcher will manipulate this variable to observe its effect on mental health.
- **Dependent variable**
  - *Mental health status* (e.g., measured using a standardized questionnaire on anxiety or depression).
  - **Why it's dependent:** This is the outcome the researcher measures to see if it changes based on exercise frequency.
- **Control variables**
  - *Age, gender, and diet* of participants.
  - **Why they're controlled?**: These factors are kept constant to ensure the results are solely due to the exercise variable and not influenced by other external factors.

1.8 OBJECTIVES OF THE RESEARCH PROBLEM

The objectives of a research problem outline the specific goals the researcher aims to achieve. These objectives guide the research process and help in designing the study. They should be:

1.8.1 Specific

Specific refers to the clarity and precision of a research problem or objective [13]. A specific research problem clearly defines the focus, ensuring that the scope of the study is narrow enough to be addressed effectively. It avoids vagueness and generalities, making the research goals clear and actionable.

Example of specific

Research problem

"The effect of technology on education."

- **Vague:** This is too broad and unclear, as it doesn't specify what aspect of technology or education is being studied.

Specific

"The effect of interactive whiteboards on student engagement in middle school classrooms."

- **Why it's specific?**

- It clearly defines the type of technology (interactive whiteboards).
- It specifies the target population (middle school students).
- It focuses on a particular outcome (student engagement).

1.8.2 Measurable

Measurable refers to the ability to quantify or assess the outcomes of a research study. A research problem or objective should be framed in a way that allows the researcher to collect data that can be measured, counted, or evaluated in a clear and consistent manner [14].

Example of measurable

Research problem

"The effect of sleep on productivity."

- **Non-measurable:** This is vague because "productivity" and "sleep" are not clearly defined in terms of measurable units.

Measurables

"The effect of 8 hours of sleep on the number of tasks completed by employees in a 4-hour work period."

- **Why it's measurable?**
  - The amount of sleep is defined (8 hours).
  - Productivity is measured by the number of tasks completed, a quantifiable outcome.
  - This allows for data collection and analysis to determine the exact effect.

By framing the research problem in measurable terms, it becomes easier to collect and analyze data, leading to more concrete and actionable results.

1.8.3 Achievable

Achievable refers to the feasibility of a research problem or objective, ensuring that it can be realistically accomplished within the available time, resources, and expertise [15]. An achievable research problem is one that can be tackled with the tools, methods, and data that are accessible to the researcher.

Example of achievable

Research problem

"The effect of climate change on global agricultural production over the next 50 years."

- **Unachievable**

- This is an extremely broad and long-term study, requiring extensive global data and forecasting models that may not be feasible for a single researcher or within the time frame of a typical research project.
- **Why it's unachievable?**: The scale and time span are beyond the researcher's control, and the resources needed are too extensive.

- **Achievable**

"The effect of temperature rise on crop yield in the Midwest U.S. over the past 10 years."

- **Why it's achievable?**
  - The time frame is more reasonable (10 years), and the geographical focus (Midwest US) is more specific and manageable.
  - The study can be done with available historical data on temperature and crop yields, making it feasible with existing resources and methods.
  - The scope is narrow enough to allow for meaningful analysis and conclusions within the researcher's capabilities.

By defining an achievable research problem, the study remains practical, focused, and within the limits of available resources and expertise.

1.8.4 Relevant

Relevant refers to the importance and applicability of a research problem or objective in addressing a current issue, filling a gap in existing knowledge, or contributing to the field. A relevant research problem is one that aligns with the needs of the field, has practical implications, and addresses significant questions or challenges [16].

Example of relevant

Research problem

"The effect of smartphone use on sleep patterns in teenagers."

- **Relevant**
  - This research is highly relevant because it addresses a significant, contemporary issue that affects many teenagers. Excessive smartphone use has been linked to poor sleep quality, and understanding this relationship could help improve health recommendations and interventions.
  - **Why it's relevant?**
    - The topic is timely and aligns with growing concerns about the impact of technology on health.
    - It has practical implications for improving sleep habits and mental health in adolescents.

- **Irrelevant**

"The impact of medieval architecture on modern interior design trends."
  - **Why it's irrelevant?**
    - While it might be an interesting topic, it's less directly applicable to current real-world issues, such as technology use or public health, and doesn't contribute significantly to pressing contemporary challenges.

By ensuring a research problem is relevant, the study can make a meaningful contribution to solving real-world problems or advancing knowledge in a specific field.

1.8.5 Time-Bound

Time-bound refers to setting a clear and realistic time frame for completing a research study or achieving its objectives [17]. A time-bound research problem or objective includes specific deadlines or milestones, ensuring that the study stays focused and is completed within a defined period. This helps manage the research process and ensures that results are timely and relevant.

Example of time-bound

Research problem

"The effect of exercise on weight loss."

- **Not time-bound**
  - This is a general statement without a clear time frame, making it open-ended and lacking focus. It doesn't specify when or over how long the exercise will be measured or the weight loss observed.

Time-bound

"The effect of 30 minutes of daily aerobic exercise on weight loss in adults over a 3-month period."

- **Why it's time-bound?**
  - The research problem specifies a clear time frame (3 months), providing a defined period for data collection and analysis. This helps in managing the study and ensures that results are obtained within a specific, practical time frame.

By making the research problem time-bound, the study becomes focused and feasible and ensures that progress can be tracked within a defined period.

1.9 APPROACHES TO INVESTIGATE RESEARCH PROBLEMS

When investigating a research problem, the approach refers to the overall strategy or methodology used to explore the issue, gather data, and analyze results. The common approaches include the following [18]

1.9.1 Quantitative approach

Quantitative approach is a research methodology that collects and analyzes numerical data to identify patterns, relationships, or trends. This approach is often used to test hypotheses, measure variables, and draw generalizable conclusions. It relies on structured tools such as surveys, experiments, or statistical analysis to produce objective and replicable results.

Key features of the quantitative approach

1. **Objective measurement:** Data is collected and analyzed in a way that minimizes subjectivity.
2. **Numerical data:** The focus is on quantifiable information, such as percentages, frequencies, or statistical measures.
3. **Hypothesis testing:** Often involves testing a specific hypothesis or research question.
4. **Large Sample size:** Data is typically collected from a large number of participants to ensure reliability and generalizability.
5. **Statistical Analysis:** Uses statistical tools to interpret data, identify patterns, and make predictions.

Methods used

1. **Surveys/questionnaires:** Closed-ended questions that generate numerical responses.
   - **Example:** Measuring customer satisfaction on a scale of 1 to 10.
2. **Experiments:** Controlled settings to manipulate variables and observe effects.
   - **Example:** Testing the effect of a new drug on blood pressure.
3. **Observational Studies:** Systematic recording of measurable behaviours or phenomena.
   - **Example:** Counting the number of cars passing through an intersection.

Advantages

- Produces objective, reliable, and replicable results.
- Enables analysis of large datasets for broader generalizations.
- Facilitates hypothesis testing and statistical modelling.

Example

Research problem

"Does regular aerobic exercise reduce blood pressure in adults?"

- **Independent variable:** Frequency of aerobic exercise (e.g., 3 times per week).
- **Dependent variable:** Blood pressure levels (measured in mmHg).

- **Method:** Conduct an experiment where one group performs aerobic exercise while another does not, and measure their blood pressure after 3 months.
- **Analysis:** Use statistical tests (e.g., t-test) to determine if exercise significantly reduces blood pressure.

1.9.2 Qualitative approach

Qualitative approach refers to a research methodology focused on exploring and understanding human experiences, behaviours, and perceptions [18]. Unlike the quantitative approach, which relies on numerical data, the qualitative approach collects non-numerical data, such as words, images, or observations, to gain deeper insights into complex phenomena.

Key features of the qualitative approach

1. **Exploratory nature:** It seeks to explore meanings, patterns, and experiences rather than test hypotheses.
2. **Non-numerical data:** Uses data like interviews, focus group discussions, observations, or textual content.
3. **Subjective interpretation:** Relies on the researcher's interpretation to understand the context and meaning of the data.
4. **Small sample size:** Often focuses on a smaller, targeted group for in-depth analysis.
5. **Flexible and open-ended:** Data collection methods are adaptable and often evolve during the study.

Methods used

1. **Interviews:** One-on-one or group conversations to explore participants' thoughts, feelings, and experiences.
   - **Example:** Asking teachers how they adapt their teaching styles to online learning.
2. **Focus groups:** Group discussions to gather diverse perspectives on a specific topic.
   - **Example:** Discussing consumer opinions about a new product design.
3. **Observations:** Watching participants in their natural environment understand behaviours or interactions.
   - **Example:** Observing how employees collaborate in an open office space.
4. **Document analysis:** Analyzing written or visual content to extract meaning.
   - **Example:** Studying social media posts to understand public opinion on climate change.

Advantages

- Provides rich, detailed insights into complex or nuanced topics.
- Captures human emotions, perceptions, and social contexts.
- Useful for exploring new or under-researched areas.

Example

Research problem

"How do patients feel about telemedicine services during the COVID-19 pandemic?"

- **Data collection:** Conduct semi-structured interviews with patients.
- **Analysis:** Use thematic analysis to identify recurring themes, such as convenience, privacy concerns, or technical challenges.
- **Outcome:** Develop a deeper understanding of patient satisfaction and concerns with telemedicine.

1.9.3 Mixed methods

Mixed methods is a research approach that combines quantitative and qualitative methods to provide a more comprehensive understanding of a research problem [18]. By integrating numerical data with in-depth insights, mixed methods capitalize on the strengths of both approaches and address their individual limitations.

Key features of mixed methods

1. **Integration of data:** Combines numerical data (quantitative) with detailed, descriptive data (qualitative).
2. **Complementary insights:** Quantitative data provides generalizable trends, while qualitative data offers depth and context.
3. **Sequential or concurrent use:** Methods can be used sequentially (one after the other) or concurrently (at the same time).
4. **Flexible design:** Adapts to the research problem's complexity, allowing researchers to address diverse aspects.

When to use mixed methods?

- When neither quantitative nor qualitative methods alone can fully address the research problem.
- To validate and enrich findings from one method with insights from the other.
- To explore a phenomenon (qualitative) and then measure it (quantitative) or vice versa.

Types of mixed methods designs

1. **Explanatory sequential:** Quantitative data is collected and analyzed first, followed by qualitative data to explain or elaborate on the findings.
   - **Example:** Surveying employees about job satisfaction and then conducting interviews to explore the reasons behind their responses.
2. **Exploratory sequential:** Qualitative data is collected first to explore a phenomenon, followed by quantitative data to test or generalize findings.
   - **Example:** Conducting focus groups to identify key themes and then creating a survey to measure those themes in a larger population.
3. **Concurrent triangulation:** Both methods are used simultaneously, and the findings are compared to validate or complement each other.
   - **Example:** Observing classroom interactions while also collecting student feedback through surveys.

Advantages

- Provides a fuller understanding of the research problem.
- Balances the strengths and limitations of quantitative and qualitative methods.
- Enhances the reliability and validity of results through data triangulation.

Example

Research problem

"What is the impact of remote work on employee productivity and well-being?"

1. **Quantitative:** Conduct a survey to measure productivity levels and well-being indicators (e.g., stress levels and work hours).
2. **Qualitative:** Conduct interviews with employees to explore their experiences, challenges, and coping mechanisms during remote work.
3. **Integration:** Use the qualitative findings to explain trends in the quantitative data (e.g., why productivity increased or decreased).

By using both **numbers and narratives, mixed methods** provide a richer, more nuanced perspective, making it particularly effective for complex research problems.

1.10 DATA COLLECTION FOR RESEARCH PROBLEM

Data collection for a research problem refers to the systematic process of gathering information or evidence to answer a research question, test hypotheses, or address objectives [19]. The method of data collection depends on the nature of the research problem, the type of data needed, and the chosen research approach (quantitative, qualitative, or mixed methods).

Types of data

1. **Primary data**

   Primary data refers to data collected directly by the researcher for a specific purpose or research study. It is original, first-hand information gathered from sources such as participants, experiments, or observations, tailored to address the research problem.

Characteristics of primary data

- **Originality:** It is collected first-hand, making it unique to the specific research study.
- **Specificity:** Directly related to the research problem or objective.
- **Control:** The researcher determines how and when the data is collected.
- **Timeliness:** Reflects current conditions, behaviours, or trends.

Methods of collecting primary data

- **Surveys/questionnaires**
  - **Description:** Structured forms with questions for participants.
  - **Example:** A survey to measure customer satisfaction with a product.
- **Interviews**
  - **Description:** One-on-one or group discussions to gather in-depth insights.
  - **Example:** Interviewing employees about workplace challenges.
- **Observations**
  - **Description:** Watching and recording behaviours or events in their natural setting.
  - **Example:** Observing traffic flow at an intersection to design a new traffic signal.
- **Experiments**
  - **Description:** Controlled studies where variables are manipulated to observe effects.
  - **Example:** Testing a new teaching method's impact on student performance.
- **Focus groups**
  - **Description:** Facilitated discussions among a small group to explore perceptions.
  - **Example:** Discussing consumer reactions to a new advertising campaign.

- Case studies
  - **Description:** In-depth exploration of a single instance, group, or event.
  - **Example:** Studying the effects of remote work in a specific organization.

Advantages

- **Specific and relevant:** Directly tailored to the research problem.
- **Up-to-date:** Reflects current data, avoiding outdated information.
- **Control over quality:** The researcher ensures data accuracy and reliability.

Disadvantages

- **Time-consuming:** Data collection can be lengthy and labour-intensive.
- **Costly:** Requires resources like travel, equipment, or participant incentives.
- **Limited Scope:** Often focuses on a smaller sample, which may reduce generalizability.

Example

Research problem

"What factors influence customer loyalty in e-commerce?"

- **Method:** Conduct an online survey with customers who frequently shop online.
- **Primary data collected:** Responses about product quality, pricing, delivery times, and customer service experiences.
- **Purpose:** Understand which factors most significantly affect customer loyalty.

Primary data is crucial for research requiring specific, first-hand information and is highly valued for its relevance and authenticity.

2. **Secondary data**

Secondary data refers to data that has already been collected, analyzed, and published by someone else for a different purpose. Researchers use this pre-existing data to support their studies, saving time and resources.

Characteristics of secondary data:

- **Pre-existing:** Collected and documented by other individuals, organizations, or institutions.
- **Easily accessible:** Available through sources like government reports, academic publications, and online databases.

- **Cost-effective:** Does not require primary data collection efforts.
- **Broad scope:** Often covers larger populations, longer timeframes, or historical perspectives.

Sources of Secondary Data

- **Government publications**
  - **Examples:** Census reports, economic surveys, and health statistics.
- **Academic research**
  - **Examples:** Peer-reviewed journals, theses, and conference proceedings.
- **Organizational records**
  - **Examples:** Annual reports, financial statements, and customer data.
- **Public databases**
  - **Examples:** World Bank, United Nations, WHO, or online repositories.
- **Media and internet**
  - **Examples:** News articles, blogs, or social media analytics.

Advantages

- **Time-saving:** No need to conduct surveys, experiments, or interviews.
- **Cost-effective:** Eliminates expenses related to primary data collection.
- **Large coverage:** Provides access to data from broader populations or regions.
- **Historical insight:** Offers long-term data trends and past records.

Disadvantages

- **Limited relevance:** Data may not perfectly align with the research problem.
- **Potential bias:** The data collection methods or intentions of the original source may introduce bias.
- **Outdated information:** Secondary data might not reflect current conditions.
- **Lack of control:** Researchers have no influence over data quality or collection methods.

Example

Research problem

"How has urban population growth affected air quality in major cities?"

- **Secondary data sources**
  - Population growth data from census reports.
  - Air quality indices from government environmental agencies.
  - Historical climate reports from international organizations.
- **Usage:** Analyze correlations between urbanization trends and air quality changes without conducting primary data collection.

Secondary data is an efficient and cost-effective resource, especially for preliminary research, comparative studies, or exploring trends over time. However, researchers must critically evaluate its relevance, accuracy, and credibility before use.

1.10.1 Data collection methods

1. **Quantitative data collection**
   - Focuses on numerical data that can be measured and analyzed statistically [20].

Methods

- **Surveys/questionnaires:** Standardized tools with closed-ended questions to gather responses from large populations.
  - **Example:** Collecting customer satisfaction scores.
- **Experiments:** Controlled studies to observe the effect of manipulating variables.
  - **Example:** Testing a new fertilizer's effect on crop yield.
- **Observations:** Counting or measuring specific behaviours or events.
  - **Example:** Recording the number of cars passing through a toll gate.

2. **Qualitative Data Collection**
   - Focuses on non-numerical, descriptive data to explore ideas, behaviours, or experiences.

Methods

- **Interviews:** Open-ended conversations to gather in-depth insights.
  - **Example:** Interviewing teachers about challenges in online education.
- **Focus groups:** Group discussions to explore collective views.
  - **Example:** Discussing consumer preferences for a new product.
- **Observations:** Watching interactions or events in natural settings.
  - **Example:** Observing patient-doctor interactions in a clinics.

3. **Mixed methods data collection**
   - Combines quantitative and qualitative methods for a comprehensive approach.

- **Example:** Conducting a survey to measure employee stress levels (quantitative) and interviews to understand their experiences (qualitative).

1.10.2 Steps in data collection

1. **Define objectives:** Clearly identify what information is needed and why.
2. **Select methods:** Choose the most appropriate method(s) based on the research problem.
3. **Develop tools:** Design surveys, interview guides, or observation checklists.
4. **Pilot testing:** Test tools on a small sample to ensure clarity and reliability.
5. **Collect data:** Execute the data collection process while ensuring accuracy and consistency.
6. **Record and store:** Safeguard collected data for analysis, maintaining ethical standards.

Key considerations

- **Validity:** Ensure the data accurately represents the phenomenon being studied.
- **Reliability:** Ensure consistency in data collection across different times or groups.
- **Ethics:** Obtain informed consent, maintain confidentiality, and respect participants' rights.
- **Feasibility:** Choose methods that are practical and within the study's budget and timeline.

Example

Research problem

"How does social media influence the academic performance of college students?"

- **Quantitative:** Use a survey to measure time spent on social media and corresponding GPA scores.
- **Qualitative:** Conduct interviews to explore students' perceptions of how social media affects their study habits.
- **Mixed Methods:** Combine both to validate survey results with detailed narratives from interviews.

Effective data collection ensures that the research findings are credible, actionable, and aligned with the research problem.

1.11 ANALYSIS FOR RESEARCH PROBLEM

Analysis for a research problem is the systematic process of examining, organizing, and interpreting collected data to address the research question or objectives [21]. It helps researchers extract meaningful insights, validate hypotheses, and draw conclusions that contribute to solving the research problem.

Key objectives of analysis

1. **Understand patterns and relationships:** Identify trends, correlations, or causal links in the data.
2. **Answer research questions:** Address the problem and objectives defined at the outset.
3. **Support decision-making:** Provide evidence-based insights for recommendations or further action.

Types of analysis

1. **Quantitative analysis**
   - Focuses on numerical data.
   - Uses statistical methods like regression, correlation, and hypothesis testing.
   - **Example:** Determining the effect of study hours on exam scores using statistical correlation.
2. **Qualitative analysis**
   - Focuses on non-numerical data such as text, interviews, or observations.
   - Uses methods such as thematic analysis or content analysis to identify recurring themes or insights.
   - **Example:** Analyzing employee interviews to understand job satisfaction factors.
3. **Mixed-methods analysis**
   - Combines both quantitative and qualitative techniques.
   - **Example:** Using surveys to quantify customer satisfaction and interviews to explore reasons behind their scores.

Steps in data analysis

1. **Data preparation**
   - Organize data: Ensure it is structured and ready for analysis.
   - Clean data: Address missing values, remove outliers, and check for inconsistencies.
2. **Descriptive analysis**
   - Summarize data using measures like mean, median, mode, and standard deviation.

- Visualize data through graphs, charts, or tables.

3. **Exploratory analysis**
   - Explore potential relationships, trends, or anomalies.
   - Use visual tools like scatter plots or heatmaps for insight.
4. **Inferential analysis**
   - Apply statistical methods to generalize findings from a sample to a larger population.
   - Test hypotheses to determine statistical significance.
5. **Interpretation**
   - Relate findings to the research objectives.
   - Highlight key insights and their implications.
6. **Validation**
   - Cross-check findings to ensure reliability and validity.
   - Compare results with existing literature or data sources.
7. **Conclusion and Reporting**
   - Summarize findings in relation to the research problem.
   - Present insights using visual aids and concise narratives.

Example

Research problem

"How does social media usage affect academic performance among college students?"

- **Data Collected**
  - Survey responses about daily social media usage and GPA scores.
  - Interviews with students about their study habits.
- **Analysis**
  - **Quantitative:** Use regression analysis to determine the relationship between social media usage and GPA.
  - **Qualitative:** Thematic analysis of interviews to identify common behaviours influencing study time.
- **Findings**
  - Students with higher social media usage tend to have lower GPAs.
  - Key reasons include procrastination and distractions during study sessions.

1.12 INTERPRETATION FOR RESEARCH PROBLEM

Interpretation for a research problem is the process of making sense of the analyzed data by linking findings to the research objectives, hypotheses, or broader context [22]. It involves explaining what the results mean, identifying implications, and providing insights that address the research problem.

Key aspects of interpretation

1. **Connecting findings to objectives**
   - Relate the results directly to the research questions or goals.
   - Determine whether the findings support or contradict the hypotheses.
2. **Explaining relationships and patterns**
   - Highlight trends, correlations, or causations found during analysis.
   - Provide a logical explanation for observed phenomena.
3. **Evaluating implications**
   - Discuss the broader impact of the findings on the field or real-world applications.
   - Suggest how the results contribute to knowledge or practice.
4. **Acknowledging limitations**
   - Identify any constraints or biases in the study that might affect the interpretation.
   - Suggest areas for further research.
5. **Contextualizing results**
   - Compare findings with existing literature to validate or challenge prior knowledge.
   - Place the results within the socioeconomic, cultural, or scientific context.

Steps in interpretation

1. **Summarize key findings**
   - Condense the main results into clear, concise points.
2. **Link to research objectives**
   - Address how each finding answers the research question or supports the hypothesis.
3. **Provide explanations**
   - Offer reasons for the observed outcomes based on theory, prior studies, or logical reasoning.
4. **Discuss implications**
   - Explore the practical, theoretical, or policy-related consequences of the results.
5. **Highlight unanswered questions**
   - Point out gaps or unexpected results that warrant further investigation.

Example

Research Problem

"How does remote work affect employee productivity in IT companies?"

- **Findings**

- Employees working remotely reported a 15% increase in productivity.
- Key factors included fewer distractions, flexible hours, and reduced commute times.
- Some challenges, such as isolation and communication issues, were also noted.

- **Interpretation**
  - The increase in productivity aligns with theories of autonomy improving performance.
  - Flexible hours allow employees to work during their most productive times, enhancing output.
  - Communication challenges suggest a need for improved virtual collaboration tools and strategies.
  - Comparison with prior studies shows consistent results, validating the reliability of the findings.
- **Implications**
  - Companies can consider hybrid models to balance productivity gains and employee well-being.
  - Investment in virtual tools and regular team-building activities may mitigate challenges.

1.13 NECESSARY INSTRUMENTATIONS FOR THE RESEARCH PROBLEM

Necessary instrumentations for a research problem refers to the tools, devices, software, or methodologies required to collect, measure, analyze, and interpret data [23]. Selecting appropriate instrumentation ensures the accuracy, reliability, and validity of the research findings.

Key considerations for instrumentation

1. **Relevance**
   - Ensure the tools align with the research objectives and data collection needs.
2. **Accuracy and precision**
   - Select instruments capable of producing reliable and consistent results.
3. **Feasibility**
   - Consider the availability, cost, and ease of use of the instruments.
4. **Standardization**
   - Use validated and widely accepted tools to maintain credibility.
5. **Ethical compliance**
   - Ensure the instrumentation does not violate ethical guidelines, especially for human or animal studies.

Types of instrumentations

1. **Data collection tools**
    - **Surveys and questionnaires:** For quantitative or qualitative data.
    - **Interview protocols:** For collecting in-depth insights.
    - **Observation checklists:** For structured or semi-structured observations.
2. **Measurement devices**
    - **Scientific instruments:** Thermometers, spectrometers, and microscopes for physical or chemical research.
    - **Sensors:** Temperature, pressure, or motion sensors for engineering or environmental studies.
    - **Psychometric tools:** Scales or tests for measuring psychological variables.
3. **Software and analytical tools**
    - **Statistical software:** SPSS, R, and Python for quantitative data analysis.
    - **Qualitative tools:** NVivo, ATLAS.ti for thematic or content analysis.
    - **Simulation tools:** MATLAB and ANSYS for modelling and simulations.
4. **Experimental setups**
    - **Laboratory equipment:** Test rigs, wind tunnels, or bioreactors.
    - **Prototyping tools:** 3D printers or CAD software for engineering research.
5. **Recording and storage tools**
    - **Audio/video recorders:** For interviews or observations.
    - **Data loggers:** For continuous monitoring of environmental or experimental variables.

Example

Research problem

"How does soil moisture content affect plant growth in controlled environments?"

- **Necessary instrumentation**
    1. **Soil moisture sensors:** To measure and monitor moisture levels.
    2. **Growth chambers:** For maintaining controlled environmental conditions.
    3. **Data loggers:** To record temperature, humidity, and moisture over time.
    4. **Measurement tools:** Calipers and scales to assess plant growth metrics like height and biomass.
    5. **Statistical software:** SPSS for analyzing correlations between soil moisture and plant growth.

1.14 SUMMARY

This chapter delves into the fundamental aspects of defining and addressing a research problem, starting with an introduction that highlights its centrality in any study. The meaning of a research problem is clarified, emphasizing its role as the foundation that shapes the direction, scope, and objectives of the research. The role of a research problem in a study is explored, underscoring how it determines the relevance and impact of the findings. Various sources of research problems, such as practical issues, theoretical gaps, and societal needs, are discussed, alongside types of research problems, including descriptive, exploratory, and causal.

The chapter outlines the criteria of a good research problem, such as clarity, feasibility, significance, and alignment with the researcher's expertise. It also identifies common errors in selecting a research problem, including lack of clarity, overly ambitious scope, repetition of previous studies, irrelevance to the field, personal bias, insufficient background research, overdependence on available data, misalignment with the researcher's expertise, ignoring feasibility, and overlooking ethical considerations. To mitigate these errors, strategies like conducting thorough literature reviews, aligning with current trends, ensuring clarity and feasibility, consulting mentors, and evaluating the problem's significance are detailed.

The scope and objectives of a research problem are discussed with specific attention to defining geographical limits, time frames, population or sample, and variables. The chapter also emphasizes the importance of setting smart objectives specific, measurable, achievable, relevant, and time-bound to ensure focused and actionable research goals.

Approaches for investigating research problems include the quantitative approach for statistical analysis, the qualitative approach for in-depth understanding, and mixed methods that combine both for comprehensive insights. The chapter further explores data collection methods, distinguishing between primary and secondary data, and provides steps for effective data collection.

Finally, the chapter discusses analysis and interpretation, emphasizing the need to derive meaningful insights from data while linking findings to the research objectives. The role of necessary instrumentation, including tools, software, and methodologies for data collection and analysis, is highlighted as essential for ensuring accuracy and validity in research. Together, these components provide a structured framework for understanding, formulating, and addressing a research problem effectively.

REVIEW QUESTIONS

1. What is a research problem, and why is it essential in a study?
2. How would you define a research problem in your own words?

3. Why the research problem is considered the foundation of any study?
4. What are some common sources of research problems? Provide examples.
5. Distinguish between different types of research problems with examples.
6. What are the key characteristics of a well-defined research problem?
7. How does feasibility play a role in evaluating a research problem?
8. What challenges arise from a lack of clarity in defining a research problem?
9. How can overly ambitious scope hinder the success of a research project?
10. Why is repetition of previous studies often discouraged?
11. Discuss how personal bias can negatively influence the selection of a research problem.
12. What are the consequences of ignoring feasibility and ethical considerations in research?
13. Why is conducting a thorough literature review crucial in avoiding errors?
14. How can aligning a research problem with current trends and needs enhance its relevance?
15. What role do mentors or experts play in refining a research problem?
16. How do geographical limits and time frames influence the scope of a research problem?
17. Why is defining variables important when framing research objectives?
18. How does each component of smart contribute to a well-defined research problem?
19. Compare and contrast quantitative, qualitative, and mixed-method approaches to solving research problems.
20. In what scenarios would a mixed-method approach be most appropriate?
21. What are the key differences between primary and secondary data?
22. Describe the steps involved in collecting data for a research problem.
23. How does data analysis transform raw data into meaningful insights?
24. Why is interpretation crucial in the research process?
25. How can necessary instrumentation influence the quality of a study?
26. What factors should be considered when selecting tools for data collection and analysis?

Chapter 2

Mastering research methodology and communication

2.1 INTRODUCTION

Research methodology and communication are critical pillars of any academic or professional endeavour. The ability to systematically investigate a research question, analyze findings, and effectively convey results defines the success of any project. Mastery in these areas not only ensures scientific rigour but also enhances the impact and relevance of the research.

This chapter explores the foundational aspects of research methodology and communication, providing insights into effective literature studies, plagiarism avoidance, adherence to research ethics, and technical writing. Additionally, it delves into practical guidance for developing a research proposal, formatting it appropriately, and presenting it to a review committee.

By integrating theoretical knowledge with actionable techniques, this chapter aims to equip readers with the tools necessary to navigate the complexities of research and communication. Whether preparing a technical report, drafting a research paper, or defending a proposal, the skills covered here are indispensable for achieving excellence in the academic and professional realms.

2.2 EFFECTIVE LITERATURE STUDIES APPROACHES

Effective literature studies are the foundation of sound research. Begin by identifying relevant databases and journals specific to your field. Use advanced search techniques, such as Boolean operators, to refine your results [24]. Organize the literature using tools like reference management software (e.g., Mendeley and Zotero). Critically evaluate sources by assessing their credibility, relevance, and contribution to your research area. Aim to identify gaps in the literature to position your research effectively [25].

Effective literature studies form the backbone of robust research. They involve systematically reviewing existing knowledge to identify gaps, establish context, and build a strong foundation for new investigations. Key approaches include.

DOI: 10.1201/9781003485827-2

2.2.1 Define objectives clearly

Define objectives clearly refers to the process of setting specific and well-defined goals for a literature review or research task. It ensures that the researcher remains focused and purposeful, avoiding unnecessary detours into irrelevant material [26]. By establishing clear objectives, researchers can determine what information is essential, identify gaps in the existing knowledge, and align their efforts with the overall aims of their study.

For example, objectives might include identifying theoretical frameworks, comparing methodologies, or understanding historical trends within a field. This clarity simplifies the selection of relevant sources and guides the structure of the review, enhancing its overall effectiveness.

2.2.2 Use diverse sources

Use diverse sources emphasizes the importance of consulting a wide range of materials to ensure a comprehensive understanding of the research topic. Relying solely on a single type of source, such as journal articles, may limit perspectives and miss critical insights [27]. A diverse array of sources provides a richer context and a balanced view of the subject.

Examples of diverse sources include:

1. **Books** for foundational knowledge and theoretical frameworks.
2. **Peer-reviewed journals** for the latest research and validated findings.
3. **Conference proceedings** for cutting-edge developments and emerging trends.
4. **Theses and dissertations** for in-depth studies on niche topics.
5. **Credible online resources** such as government reports or industry whitepapers.

By using multiple types of sources, researchers can cross-verify information, explore different methodologies, and identify broader patterns, resulting in a well-rounded and robust literature review.

2.2.3 Employ advanced search techniques

Employ advanced search techniques refers to the use of sophisticated methods to locate high-quality and relevant information efficiently during literature studies [28]. These techniques go beyond simple keyword searches, enabling researchers to refine and narrow their search results for precision and depth.

Examples include:

1. **Boolean operators:** Using "AND," "OR," and "NOT" to combine or exclude keywords (e.g., "smart materials AND drones").

2. **Wildcards and truncation:** Using symbols like * or ? to find variations of a word (e.g., "model*" retrieves "modelling," and "models").
3. **Filters and limits:** Applying filters for publication year, document type, or language to focus on the most relevant materials.
4. **Database-specific features:** Utilizing advanced tools in platforms like Scopus, PubMed, or IEEE Xplore for citation tracking or topic clustering.
5. **Search alerts:** Setting up notifications for new publications on a specific topic.

These techniques save time, improve the quality of the results, and help in identifying hidden but critical resources for the study.

2.2.4 Critically evaluate sources

Critically evaluate sources involves assessing the credibility, relevance, and quality of the information obtained during literature studies [29]. Not all sources are equally reliable or applicable, and critical evaluation ensures that only the most trustworthy and pertinent materials inform the research.

Key aspects of evaluation include:

1. **Authority**
   Authority in the context of evaluating sources refers to assessing the expertise and credibility of the author or organization responsible for the information. Reliable sources are typically created by individuals or institutions with recognized qualifications and a proven track record in the relevant field.

   Key considerations for evaluating authority

   - **Author credentials**
     Look for academic degrees, professional experience, or publications that demonstrate expertise
     Example: A research paper on aerospace engineering authored by a professor with a PhD in Mechanical Engineering is more authoritative than one by an unverified writer.
   - **Institutional affiliation**
     Check if the author is associated with a reputable university, research organization, or industry leader.
     Example: An article from NASA on drone technology carries more authority than a blog post from an unknown website.
   - **Publication source**
     Ensure the work is published in a recognized, peer-reviewed journal, book, or conference proceedings.
     Example: Studies published in *Nature* or *IEEE Xplore* are highly credible.

- **Citations and references**
  Evaluate whether the author's work is frequently cited by others, indicating influence and recognition in the field.
- **Professional reputation**
  Look for memberships in professional organizations or awards that signify the author's standing in their domain.

2. **Accuracy**
   Accuracy refers to the reliability, correctness, and precision of the information presented in a source. It ensures that the data or arguments used in research are factual, verifiable, and consistent with established knowledge in the field.

   Key considerations for evaluating accuracy

   - **Fact-checking:** Verify claims, statistics, or data against multiple reputable sources.
     Example: A paper stating "drone adoption has grown by 300% in the last five years" should provide evidence from industry reports or official statistics.
   - **References and citations:** Check if the source cites reliable references to support its claims. A lack of citations or use of dubious references reduces accuracy.
     Example: A study on smart materials in drones that cites peer-reviewed journals is more accurate than one referencing anonymous blogs.
   - **Methodology transparency:** For research papers, evaluate whether the methods used to gather and analyze data are clearly described and follow accepted scientific standards.
     Example: An experimental study should detail its sample size, variables, and procedures.
   - **Consistency:** Ensure the information aligns with other authoritative sources in the field. Discrepancies may indicate errors or biases.
   - **Objective presentation:** Assess if the information is presented without exaggeration, distortion, or omission of critical details.
   - **Updates:** Confirm that the source reflects the latest knowledge or technological advancements, particularly in rapidly evolving fields like drones or AI.

3. **Relevance**
   Relevance refers to how closely a source aligns with the specific research problem, objectives, or context. It ensures that the information contributes directly to addressing the research questions and supports the study's goals.

   Key considerations for evaluating relevance:

   - **Alignment with research objectives**
     Determine whether the source provides insights or data that address your research questions or hypotheses.

Example: If studying the impact of drone technology on logistics, a source discussing drones in agriculture might be less relevant unless it provides transferable insights.

- **Context-specific information**
  Assess whether the source fits the geographical, cultural, or industry context of your research.
  Example: A study on urban drone delivery in Europe may not fully apply to rural regions in Asia due to differing infrastructure and regulations.
- **Recency**
  Ensure the source reflects current trends, theories, or technological advancements, especially in fast-changing fields.
  Example: A 2024 report on artificial intelligence (AI) integration in drones is more relevant than one from 2015, given the rapid evolution of the technology.
- **Depth of information**
  Evaluate whether the source provides enough detail to support your arguments or fill knowledge gaps in your study.
  Example: A brief news article on drone regulations may be less relevant than a comprehensive policy analysis report.
- **Target audience match**
  Consider whether the source is written for a technical, academic, or general audience and aligns with the complexity level required for your research.
  Example: For a scholarly paper, a peer-reviewed journal is more relevant than a popular magazine article.
- **Theoretical or practical application**
  Identify whether the source contributes theoretical insights or practical applications pertinent to your study.
  Example: A case study on successful drone deployments can provide relevant practical applications for a project focused on implementation strategies.

4. **Publication source**
   Publication source refers to the platform, journal, or medium where the research material is published. Evaluating the credibility and reliability of the publication source is essential to ensure the quality of the information used in research.

Key considerations for evaluating publication source

- **Reputation of the publisher**
  Reputable publishers are often associated with rigorous peer-review processes and high-quality research.
  Example: Articles from publishers like Springer, Elsevier, or IEEE are typically more credible than unverified online blogs.

- **Type of source**
  Determine whether the source is academic, professional, or general. Academic sources, such as peer-reviewed journals, are usually more reliable.
  Example: A journal article in *Nature* is more authoritative than a post on a social media platform.
- **Peer-reviewed status**
  Peer-reviewed publications undergo a thorough evaluation by experts, ensuring the reliability of the research.
  Example: A paper in a peer-reviewed journal like *The Lancet* carries more weight than an opinion piece in a trade magazine.
- **Impact factor and indexing**
  Journals with a high impact factor and inclusion in respected indexing services like Scopus or Web of Science indicate credibility.
  Example: A study published in a journal with a high impact factor reflects wider acceptance in the research community.
- **Publisher's integrity**
  Be cautious of predatory journals or publishers that lack transparency and rigorous standards.
  Example: Journals that charge high publication fees without providing legitimate peer review are unreliable.
- **Relevance to the field**
  Use sources from journals or publishers specializing in your area of study to ensure the content is directly applicable.
  Example: For research on drone technology, journals like *Robotics and Autonomous Systems* or the *Journal of Unmanned Vehicle Systems* are ideal.
- **Publication date**
  Recent publications are more likely to contain up-to-date data and insights, especially in fast-evolving fields.
  Example: A 2024 article on advancements in AI for drones is more valuable than a 2010 publication on the same topic.
- **Author affiliation and expertise**
  Consider the qualifications and affiliations of the authors to validate the source's credibility.
  Example: Research by a university professor with extensive publications in the field is more credible than an anonymous author.

5. **Timeliness:**
   Timeliness refers to the relevance of a source's publication date concerning the research topic. In many fields, especially those involving rapidly evolving knowledge or technology, using the most up-to-date information is critical to ensure the research remains relevant and accurate.

   Key considerations for evaluating timeliness

   - **Field-specific relevance**

In fast-changing fields like AI, medicine, or drone technology, recent publications are essential.

Example: A 2023 study on drone communication systems is more relevant than a similar study from 2010 due to advancements in technology.

- **Historical vs. current relevance**
  While recent sources are crucial, older sources may still be valuable for historical context or foundational theories.
  Example: A classic paper from the 1970s on the Bernoulli principle remains relevant for fundamental aerodynamics but would not suffice for the latest drone propulsion technologies.
- **Updates and revisions**
  Prefer updated or revised editions of books and reports, as they often incorporate the latest developments.
  Example: The second edition of a textbook on material science published in 2024 may include new research not present in the 2015 edition.
- **Publication frequency**
  Journals that publish frequently (monthly or quarterly) are more likely to provide timely information than annual publications.
  Example: A monthly journal on renewable energy is likely to have more recent studies than an annual report.
- **Project Timeline**
  Ensure that the source's timeliness aligns with the scope and objectives of your research.
  Example: A researcher studying COVID-19 vaccine efficacy in 2024 should use data and studies from the past three years to reflect the latest findings.
- **Technological advancements**
  For topics influenced by technology, prioritize the most current research to account for recent breakthroughs.
  Example: Research on autonomous drones in 2024 would require studies that include advancements in AI and sensor technologies from the last 2–3 years.

6. **Bias and objectivity**
   Bias and objectivity refer to the need for impartiality and balanced perspectives in research sources. Bias occurs when a source reflects a one-sided viewpoint or agenda, which can compromise the credibility of the research. Objectivity ensures that information is presented fairly, without distortion or undue influence.

Key considerations for evaluating bias and objectivity

- **Author's perspective**
  Assess whether the author presents a balanced argument or promotes a particular viewpoint.

Example: A study funded by a pharmaceutical company claiming its drug is superior might require scrutiny for potential bias.

- **Language and tone**
  Objective sources use neutral and factual language, while biased sources often employ emotive or persuasive language.
  Example: A scientific article stating, "The evidence strongly supports this treatment's efficacy," is more objective than one claiming, "This treatment is a miracle cure."
- **Conflict of interest**
  Investigate whether the author or publisher has affiliations that could influence the content.
  Example: A climate change study authored by a scientist working for an oil company may require careful evaluation for impartiality.
- **Presentation of evidence**
  Objective sources present data and conclusions transparently, citing multiple perspectives. Biased sources may cherry-pick evidence to support their argument.
  Example: A policy report on renewable energy that includes both benefits and challenges demonstrates objectivity, whereas one highlighting only benefits may indicate bias.
- **Publication source**
  Reputable, peer-reviewed journals or academic publishers are less likely to contain biased content compared to blogs or advocacy websites.
  Example: Articles from *Nature* or *Science* are typically more objective than those from partisan news outlets.
- **Omission of contradictory views**
  Bias can also manifest through the exclusion of counterarguments or alternative perspectives.
  Example: A research paper on drone technology that only discusses its advantages without mentioning privacy concerns may lack objectivity.
- **Funding sources**
  Examine the funding or sponsorship of the research to identify potential influences on its findings.
  Example: A report funded by a drone manufacturing company advocating for deregulated drone usage might reflect a vested interest.

Strategies to ensure objectivity in research

- Cross-check information across multiple sources to identify discrepancies.
- Favour peer-reviewed articles, government reports, and academic publications.
- Critically analyze the methodology and evidence presented in the source.

2.2.5 Organize findings logically

Organize findings logically is about structuring research data and insights in a coherent, systematic manner [30]. Logical organization helps readers easily follow the flow of ideas, identify key themes, and understand the relationships between different pieces of information.

Key principles for logical organization

2.2.5.1 *Thematic grouping*

Thematic grouping is an organizational approach where research findings or data are categorized and presented based on common themes or topics. This method is particularly useful for highlighting patterns, trends, and relationships within the data, making the information more accessible and easier to analyze.

Key features of thematic grouping

- **Focus on themes**
  Group related data under specific, clearly defined topics.
  Example: In a study on environmental conservation, themes might include "Deforestation Rates," "Wildlife Preservation," and "Community Involvement."
- **Logical flow**
  Arrange themes in a sequence that builds understanding, such as from broad to specific or simple to complex.
  Example: A report on healthcare might progress from "General Health Statistics" to "Impact of Policies" and then to "Future Recommendations."
- **Enhanced comparability**
  Grouping data thematically allows for easier comparison between different categories.
  Example: A marketing study might compare "Consumer Preferences by Age" and "Preferences by Region" as separate themes.

Steps to implement thematic grouping

- **Identify core themes**
  Review the research data to determine recurring topics or categories.
  Example: In a survey about work-from-home experiences, common themes could include "Productivity," "Work-Life Balance," and "Technological Barriers."
- **Create subsections**
  Develop headings or subheadings for each theme to clearly delineate them in your report or presentation.
  Example: Under "Productivity," include subsections like "Task Completion Rates" and "Employee Feedback."

- **Organize supporting evidence**
  Place relevant data, quotes, or examples under each theme to substantiate findings.
  Example: Under the theme "Impact of Social Media on Youth," include statistics, case studies, and expert opinions.
- **Maintain consistency**
  Use a consistent structure within each thematic section to enhance readability.
  Example: Begin each theme with an introduction, followed by data presentation, analysis, and conclusions.

Example of thematic grouping

Research topic: The effects of urbanization on local ecosystems

- **Theme 1: Habitat loss**
  - Evidence: Data on land use changes and deforestation rates.
  - Analysis: Discussion on species displacement and loss of biodiversity.
- **Theme 2: Pollution levels**
  - Evidence: Air and water quality reports.
  - Analysis: Correlation between urban development and pollution.
- **Theme 3: Climate change impact**
- Evidence: Temperature and precipitation data over decades.
- Analysis: Effects of urban heat islands on local weather patterns.

Benefits of Thematic Grouping

- Simplifies complex information by organizing it into digestible sections.
- Highlights connections and contrasts between themes.
- Facilitates clearer communication of research findings to the audience.

2.2.5.2 *Chronological order*

Chronological order is an organizational approach where information is arranged in the sequence of events or steps as they occur over time [31]. This method is particularly useful for showing progression, cause-and-effect relationships, or historical development within a topic.

Key features of chronological order

- **Time-based structure**
  Information is presented according to its timeline, starting with the earliest event and moving forward.
  Example: In a historical study of aviation, the timeline might begin with the Wright brothers' flight in 1903 and progress to modern drone technology.

- **Cause-and-effect relationships**
  Highlights how earlier events or actions influence subsequent outcomes.
  Example: In climate change research, early industrialization can be linked to rising greenhouse gas levels.
- **Step-by-step progression**
  Ideal for procedural explanations or documenting a process.
  Example: A guide to conducting an experiment might detail each phase in chronological order, from preparation to conclusion.

Steps to Implement Chronological Order

- **Identify the sequence**
  Determine the events, developments, or steps relevant to your topic and arrange them by their occurrence in time.
  Example: For a research proposal, start with problem identification, followed by a literature review, methodology, and expected outcomes.
- **Use clear time markers**
  Include dates, time periods, or phrases that indicate the order of events.
  Example: Use phrases like "In 1995," "During the 20th century," or "Subsequently."
- **Provide context**
  Explain the significance of each event or step in relation to the overall narrative or objective.
  Example: When discussing technological advancements, mention how each innovation addressed a specific challenge of its time.
- **Maintain consistency**
  Follow the same timeline throughout to avoid confusion.
  Example: Do not jump back and forth between time periods unless explicitly clarifying connections.

Example of chronological order

Research topic: Evolution of renewable energy technologies

- **Phase 1: Early innovations** (1970s–1980s)
  - Description: Initial development of solar panels and wind turbines.
  - Significance: Response to the oil crisis and rising environmental concerns.
- **Phase 2: Growth and expansion** (1990s–2000s)
  - Description: Increased investment in renewable energy and advancements in efficiency.
  - Significance: Global initiatives like the Kyoto Protocol driving adoption.

- **Phase 3: Modern era** (2010s–Present)
  - Description: Emergence of smart grids and integration of AI in energy systems.
  - Significance: Focus on sustainability and reducing carbon footprints.

Benefits of chronological order

- Clarifies progression and development over time.
- Simplifies understanding of complex sequences or histories.
- Helps audiences follow a logical flow of events or steps.

When to use chronological order

- To present historical overviews or timelines.
- To explain processes or procedures.
- To analyze trends or developments over time.

2.2.5.3 Cause-and-effect relationships

Cause-and-effect relationships explore how one event, action, or condition (the cause) leads to or influences another event or condition (the effect) [32]. This approach is widely used in research to analyze problems, identify contributing factors, and demonstrate outcomes.

Key features of cause-and-effect relationships

- **Focus on relationships**
  Establishes a logical connection between the cause and its effect.
  Example: Increased greenhouse gas emissions (cause) lead to global warming (effect).
- **Explains processes**
  Describes how and why specific outcomes occur.
  Example: Lack of sleep impairs cognitive performance by disrupting memory consolidation.
- **Multi-level analysis**
  May involve direct and indirect effects.
  Example: Industrial pollution (direct cause) not only harms local ecosystems (direct effect) but also contributes to climate change (indirect effect).

Steps to establish cause-and-effect relationships

- **Identify the cause**
  Determine the primary event, action, or condition under investigation.
  Example: The cause could be the overuse of antibiotics.
- **Identify the effect**
  Clarify the outcomes or consequences of the cause.

Example: The effect could be the emergence of antibiotic-resistant bacteria.

- **Use supporting evidence**
  Provide data, studies, or examples that demonstrate the connection.
  Example: Cite research showing how prolonged antibiotic exposure selects for resistant bacterial strains.
- **Explain the mechanism**
  Describe the process linking the cause to the effect.
  Example: Overuse of antibiotics kills susceptible bacteria, allowing resistant strains to thrive and multiply.
- **Consider other factors**
  Acknowledge any additional variables or alternative explanations.
  Example: Poor hygiene practices can also contribute to the spread of resistant bacteria.

Example of cause-and-effect relationships

Research topic: The impact of deforestation on biodiversity

- **Cause:** Deforestation due to agricultural expansion.
- **Effect:** Loss of biodiversity as habitats are destroyed.
- **Mechanism:** Clearing forests eliminates food and shelter for species, leading to population declines or extinction.
- **Indirect effects:** Loss of biodiversity affects ecosystem services like pollination and climate regulation.

Applications in research

- **Problem analysis**
  Helps identify the root causes of complex issues.
  Example: High dropout rates in schools may result from poor teacher-student engagement.
- **Predicting outcomes**
  Evaluates the potential impact of proposed solutions or interventions.
  Example: Introducing renewable energy reduces carbon emissions, mitigating climate change effects.
- **Policy formulation**
  Provides evidence for crafting effective strategies.
  Example: Restricting single-use plastics (cause) reduces ocean pollution (effect).

Benefits of using cause-and-effect relationships

- Enhances understanding of interconnected phenomena.
- Provides a logical structure for arguments and analyses.
- Identifies actionable insights to address problems.

2.2.6 Comparative analysis

Comparative analysis is a research method used to evaluate and compare two or more entities, such as concepts, theories, events, processes, or case studies [33]. It highlights similarities, differences, and patterns to derive meaningful insights or conclusions. This approach is particularly useful in identifying trends, evaluating effectiveness, and developing theoretical frameworks.

Key features of comparative analysis

- **Dual focus**
  Compares multiple subjects to understand their relationships or distinctions.
  ***Example*:** Comparing renewable energy policies in developed versus developing nations.
- **Contextual examination**
  Evaluates the entities within their respective contexts.
  ***Example*:** Comparing urban development strategies in cities with varying population densities.
- **Objective and systematic**
  Uses defined criteria to ensure fair and consistent comparisons.
  ***Example*:** Assessing the effectiveness of two teaching methods based on student performance metrics.

Steps in conducting comparative analysis

- **Define objectives**
  Clearly outline the purpose of the comparison.
  ***Example*:** To evaluate the efficiency of online and offline retail models.
- **Select subjects**
  Identify entities or variables to compare.
  ***Example*:** Compare electric and gasoline-powered vehicles.
- **Establish criteria**
  Determine the basis for comparison.
  ***Example*:** Compare vehicles based on cost, environmental impact, and performance.
- **Gather data**
  Collect relevant data for each entity using consistent methods.
  ***Example*:** Use industry reports, surveys, and test results for data on vehicle performance.
- **Analyze findings**
  Examine similarities, differences, and implications.
  ***Example*:** Electric vehicles have lower running costs but higher initial costs compared to gasoline vehicles.

- **Draw conclusions**
  Summarize the key insights and implications of the analysis.
  Example: Electric vehicles are more cost-effective and environmentally friendly in the long term.

Example of comparative analysis

Topic: Comparing the educational outcomes of online and traditional classrooms

- **Criteria for Comparison**
  - Student engagement
  - Academic performance
  - Accessibility
  - Cost
- **Findings**
  - *Engagement*: Traditional classrooms promote better face-to-face interaction.
  - *Performance*: Academic results are comparable in both models when online tools are well-designed.
  - *Accessibility*: Online learning provides greater flexibility for remote learners.
  - *Cost*: Online education is more affordable for institutions and students.
- **Conclusion**
  While traditional classrooms excel in engagement, online models offer broader accessibility and cost efficiency.

Applications of comparative analysis

- **Decision-making**
  Helps organizations choose between competing options.
  Example: Comparing cloud storage solutions for cost, security, and scalability.
- **Policy development**
  Guides policymakers by analyzing the outcomes of different approaches.
  Example: Comparing healthcare systems in countries to recommend reforms.
- **Theoretical advancement**
  Supports hypothesis testing by evaluating multiple cases or variables.
  Example: Comparing customer loyalty in different marketing strategies.

Benefits of comparative analysis

- Clarifies strengths and weaknesses of each entity.
- Highlights areas for improvement or innovation.
- Facilitates evidence-based decision-making.

2.2.7 Hierarchical structuring

Hierarchical structuring is an organizational approach where information is arranged in a top-down manner, starting with broad concepts and progressively narrowing down to more specific details [34]. This method provides a clear framework for understanding complex topics by presenting them in a logical and layered manner.

Key features of hierarchical structuring

- **Top-down approach**
  Begins with overarching themes or categories and moves towards finer details.
  Example: Starting with "Renewable Energy" and breaking it down into "Solar," "Wind," and "Hydropower."
- **Logical progression**
  Ensures that ideas flow naturally, making the content easy to follow.
  Example: Discussing general economic trends before delving into sector-specific impacts.
- **Categorization**
  Groups-related information under relevant headings or subheadings.
  Example: In a study on ecosystems, categories might include "Terrestrial," "Aquatic," and "Urban."

Steps to apply hierarchical structuring

- **Identify core themes**
  Determine the main ideas or concepts.
  Example: For a report on education, the core themes could be "Access," "Quality," and "Innovation."
- **Subdivide themes**
  Break down each theme into subcategories or components.
  Example: Under "Access," include "Geographical Barriers," "Economic Barriers," and "Digital Divide."
- **Organize logically**
  Arrange categories in order of importance or relevance.
  Example: Discuss primary challenges before addressing solutions.
- **Link-related sections**
  Show connections between categories for a cohesive narrative.
  Example: Link "Digital Divide" under "Access" to "Technology Integration" under "Innovation."

Example of hierarchical structuring

Topic: Climate change mitigation

- **Level 1: Strategies for mitigation**
  - **Level 2: Policy interventions**

    - Level 3: Carbon Taxes
    - Level 3: Renewable energy mandates
- **Level 2: Technological innovations**
    - Level 3: Carbon capture and storage
    - Level 3: Electric vehicles
- **Level 2: Community initiatives**
    - Level 3: Urban green spaces
    - Level 3: Recycling programs

Benefits of hierarchical structuring

- **Improved clarity**
  Readers can easily grasp the relationship between broad ideas and specific details.
- **Simplifies complexity**
  Breaks down intricate topics into manageable sections.
- **Facilitates navigation**
  Readers can quickly locate specific information within a structured framework.
- **Supports logical argumentation**
  Enhances the flow of reasoning, especially in research or technical writing.

Applications of hierarchical structuring

- **Research papers:** To present theories, findings, and implications in a structured format.
- **Business reports:** For organizing market analysis, strategies, and recommendations.
- **Educational content:** To teach complex subjects in a step-by-step manner.

By employing hierarchical structuring, researchers and writers can present their work in an organized, accessible, and professional manner, making it easier for audiences to engage with and understand the material.

2.2.8 Highlighting key findings

Highlighting key findings is the process of emphasizing the most critical results or insights derived from research or analysis. It ensures that the core messages are easily identifiable and accessible to the audience, enhancing the impact and clarity of the work.

Strategies for highlighting key findings

- **Summarize in bullet points**

Use concise bullet points to list the most significant findings.

Example:

1. Solar energy adoption increased by 25% in 2023.
2. Urban areas show higher recycling rates than rural regions.

- **Create visual aids**
  Use charts, graphs, or infographics to visually represent data.
  Example: A bar graph showing year-over-year growth in renewable energy usage.
- **Emphasize with formatting**
  Highlight key phrases using bold, italics, or underlining.
  Example: *The study found that 60% of respondents prefer remote work options.*
- **Use headings and subheadings**
  Clearly label sections dedicated to key findings.
  Example: Key Findings on Consumer Behaviour Trends
- **Incorporate quotes or key statements**
  Include impactful statements from the research.
  Example: "Digital literacy improved productivity by 40% in surveyed organizations."
- **Link findings to objectives**
  Directly connect findings to the research questions or objectives to reinforce relevance.
  Example: If the objective was to evaluate renewable energy adoption, highlight specific data on adoption rates.

Examples of highlighting key findings

Topic: Impact of online learning on student performance

- **Key finding 1**: Students who used interactive platforms scored 20% higher on assessments compared to those using traditional methods.
- **Key finding 2**: A total of 70% of participants reported improved time management skills.
- **Key finding 3**: Challenges included limited internet access, affecting 15% of rural participants.

Visualization: Include a pie chart showing percentages of challenges faced by participants.

Benefits of highlighting key findings

1. **Clarity**: Simplifies complex results, making them understandable at a glance.
2. **Engagement**: Draws attention to the most impactful aspects of the research.
3. **Efficiency**: Saves time for readers by focusing on essential outcomes.

4. **Improved communication:** Facilitates effective sharing of insights in presentations, reports, or discussions.

Applications of highlighting key findings

- **Research Reports:** To summarize outcomes and support conclusions.
- **Executive Summaries:** To provide decision-makers with quick insights.
- **Presentations:** To engage audiences with visually appealing and concise information.

By systematically highlighting key findings, researchers can ensure that their most important insights are communicated effectively and leave a lasting impression on their audience.

Strategies for logical presentation

- **Use headings and subheadings:** Break down findings into clear sections with descriptive titles.
- **Tables and figures:** Visual aids like charts and graphs can summarize complex data effectively.
- **Consistent formatting:** Maintain uniform style for tables, bullet points, and text for clarity.
- **Transition statements:** Use connecting phrases to explain the flow between sections.

  Example: "The next section explores the economic implications of these findings."

Example of logical organization

Research Topic: Impact of Online Education on Student Performance

- **Introduction:** Overview of online education and research objectives.
- **Findings:**
  - **Student engagement:** Analysis of attendance and participation data.
  - **Academic performance:** Comparison of grades between online and traditional classrooms.
  - **Technological challenges:** Summary of technical issues faced by students and teachers.
- **Conclusion:** Synthesis of key insights and their implications.

2.2.9 Document references

Documenting references is a critical aspect of any research endeavour, serving to acknowledge the work of others and provide a foundation for your findings [35]. Proper citation not only lends credibility to your work but also demonstrates ethical research practices and academic integrity.

Importance of documenting references

1. **Credibility:**
   Credibility in research refers to the trustworthiness and reliability of the sources cited in your work. Properly documenting credible references enhances the legitimacy of your research and strengthens the arguments you present.

   Why credibility matters

   - **Validates research:** Using authoritative and well-regarded sources supports your findings and demonstrates thorough research.
   - **Builds trust:** Credible references reassure readers of the accuracy and dependability of your work.
   - **Avoids misleading information:** Reliable sources prevent the inclusion of biased, outdated, or false data.

   Characteristics of credible sources

   Authoritative authors: Authored by experts or recognized professionals in the field.
   Peer-reviewed publications: Published in academic journals or books with rigorous review processes.
   Reputable publishers: Issued by established organizations, academic institutions, or credible publishers.
   Evidence-based content: Supported by data, experiments, or scholarly consensus.
   Timeliness: Relevant and up-to-date for the subject matter.

   How to ensure credibility

   - **Verify author credentials:** Check the qualifications and expertise of the author.
   - **Assess the publication:** Look for peer-reviewed journals or recognized publishers.
   - **Cross-check information:** Compare findings across multiple credible sources to ensure consistency.
   - **Use reliable databases:** Access research through trusted databases like PubMed, IEEE Xplore, or JSTOR.

By prioritizing credibility in your references, you reinforce the academic rigour and trustworthiness of your research, making it more persuasive and impactful.

2. **Avoiding Plagiarism**
   Plagiarism is the act of using someone else's work, ideas, or words without proper acknowledgement, which undermines the integrity of

research [36]. Avoiding plagiarism is crucial to maintaining academic honesty, building credibility, and respecting intellectual property.

Key strategies to avoid plagiarism

- **Understand plagiarism:** Recognize that copying text, paraphrasing without citation, and using ideas without acknowledgement all constitute plagiarism.
- **Cite sources properly:** Always give credit to original authors for their work, whether quoting directly or paraphrasing. Use established citation styles such as APA, MLA, or Chicago.
- **Paraphrase thoughtfully:** Rewrite ideas in your own words while preserving the original meaning, and provide proper attribution. Avoid simply changing a few words.
- **Use quotation marks:** Enclose directly quoted text in quotation marks and include a citation to the source.
- **Maintain a research log:** Keep detailed notes on the sources of your ideas and data to ensure accurate attribution during writing.
- **Leverage plagiarism detection tools:** Use software like Turnitin or Grammarly to check for unintentional plagiarism before submission.
- **Understand common knowledge:** Information that is widely known and not disputed (e.g., historical dates and widely accepted facts) does not require citation. However, any analysis or interpretation must be credited.
- **Seek permission when necessary:** For copyrighted materials, such as images or tables, ensure you have the rights or permission to use them.

Consequences of plagiarism

- **Academic penalties:** Loss of credibility, rejection of work, or academic disciplinary action.
- **Legal repercussions:** Fines or lawsuits in cases of intellectual property theft.
- **Damage to reputation:** Harm to professional and personal standing in the academic or professional community.

By following these practices, researchers can uphold ethical standards, contribute meaningfully to their field, and maintain the integrity of their work.

3. **Traceability**

   Traceability in research refers to the ability to track the origin and evolution of ideas, data, and findings throughout the research process. It ensures that every piece of information used or generated in a study can be linked back to its source, promoting transparency, reproducibility, and accountability.

Importance of traceability

- **Ensures credibility:** By providing clear links to original sources, traceability builds trust in the validity of the research.
- **Facilitates reproducibility:** Traceable data and methodologies allow other researchers to replicate studies, confirming or challenging findings.
- **Enhances accountability:** Researchers can justify their interpretations and conclusions by referencing their sources and methods.
- **Prevents misrepresentation:** Proper documentation avoids misattributing ideas or data, reducing the risk of ethical breaches like plagiarism.

Key practices for ensuring traceability

- **Maintain a research log:** Record all sources, including bibliographic details, methodologies, and datasets used during the research process.
- **Document methodologies:** Clearly outline the procedures, tools, and techniques employed to ensure others can follow or verify your process.
- **Use version control:** For collaborative projects or evolving datasets, track changes and maintain versions to understand how findings were derived.
- **Cite sources appropriately:** Ensure all referenced materials are properly cited using consistent citation styles.
- **Label data and results clearly:** Include identifiers such as timestamps, dataset versions, or experiment codes to connect data to its origin.
- **Use reference management tools:** Software like EndNote, Zotero, or Mendeley can help manage and organize references efficiently.
- **Archive raw data:** Store original datasets and documentation securely, enabling future access and verification.
- **Provide clear references in outputs:** Ensure that research papers, presentations, and reports explicitly mention all relevant sources and methodologies.

Example of traceability

In a study analyzing climate change, the researcher documents:

- The dataset used (e.g., "NOAA Global Temperature Records, Version 3.1").
- The statistical methods applied (e.g., regression analysis using R software, version 4.2.0).
- Bibliographic details for referenced studies.
- A log of modifications made to the dataset for analysis.

This level of detail ensures that others can verify, replicate, or build upon the study confidently.

By prioritizing traceability, researchers contribute to the reliability and integrity of their field, fostering a collaborative and transparent research environment.

4. **Recognition**

Recognition in research refers to the acknowledgement of the contributions of others, including ideas, data, methodologies, or collaborative efforts that have influenced or supported the study. Proper recognition fosters ethical research practices, builds trust, and enhances the collaborative spirit within the academic and professional community.

Importance of recognition

- **Ethical integrity:** Recognizing contributors ensures fairness and prevents intellectual theft or plagiarism.
- **Strengthens collaboration:** Acknowledging others' work encourages future partnerships and fosters goodwill.
- **Enhances credibility:** Proper recognition demonstrates that the researcher is aware of and respects the foundational work in their field.
- **Encourages knowledge sharing:** Acknowledgement motivates others to share their work openly, knowing their contributions will be valued.

Ways to provide recognition

- **Cite sources:** Acknowledge prior research, data, or ideas by citing them accurately in your work using standard citation formats (e.g., APA, MLA, and IEEE).
- **Acknowledge collaborators:** Mention the names of colleagues, mentors, or institutions that supported the research in an acknowledgement section.
- **Credit funding bodies:** Recognize the organizations or agencies that provide financial or material support.
- **Reference tools and software:** If specific tools, methodologies, or software were integral to the research, mention them in the methodology or acknowledgement sections.
- **Honour intellectual contributions:** Recognize key ideas or feedback received during discussions, conferences, or peer reviews.
- **Co-authorship:** Include significant contributors as co-authors on publications where applicable.

Example of recognition

In a study on renewable energy technologies:

- The researcher cites foundational papers that developed the theoretical framework.

- An acknowledgement section mentions the university's research lab and funding from a renewable energy grant.
- The research team includes collaborators as co-authors who contributed to data analysis and fieldwork.

By providing recognition, researchers uphold ethical standards, strengthen professional relationships, and contribute to the collective advancement of their field.

2.3 ANALYSIS OF LITERATURE

Analysis of literature involves critically examining existing research to understand trends, gaps, methodologies, and conclusions relevant to a specific field or research problem. This process goes beyond merely summarizing previous works and focuses on synthesizing insights to inform new research directions.

2.3.1 Steps in literature analysis

2.3.1.1 Identify core themes

Determine the central ideas and recurring concepts across various studies to understand the primary focus of research in the field [37]. Identifying core themes in literature involves extracting the central ideas and recurring patterns from multiple studies. These themes represent the fundamental topics or issues consistently explored within a field of research. Recognizing them helps establish a comprehensive understanding of the domain and guides the direction of new investigations.

Steps to identify core themes

- **Read broadly:** Review a diverse range of sources, including journal articles, books, and conference papers, to gain an overview of the field.
- **Note recurring topics:** Look for repeated keywords, concepts, or areas of focus across studies. For example, in research on renewable energy, themes like "solar efficiency," "storage solutions," and "policy impacts" may emerge.
- **Group related ideas:** Organize findings into categories based on similarity or connection. For instance, studies on environmental sustainability might group into themes such as "green technologies," "behavioural change," and "economic policies."
- **Prioritize importance:** Identify which themes are most frequently discussed or appear most significant to the field. This helps narrow down key areas for deeper analysis.
- **Align with research goals:** Ensure the themes identified are relevant to your research problem or objectives.

Importance of identifying core themes

- **Focuses research:** Pinpoints critical areas for in-depth study.
- **Reveals trends:** Highlights the evolution of ideas and emerging areas of interest.
- **Facilitates synthesis:** Helps integrate findings from diverse sources into a coherent framework.

Example

In the field of AI **ethics**, a review of the literature might reveal these core themes:

- Bias in AI algorithms.
- Privacy concerns in data usage.
- Accountability and transparency in decision-making.
- Socioeconomic impacts of automation.

By focusing on these themes, a researcher can structure their study to address gaps or build on existing work effectively. Identifying core themes ensures a strategic approach to research and contributes to meaningful advancements in the field.

2.3.1.2 Evaluate methodologies

Evaluating methodologies involves critically analyzing the research methods used in previous studies to determine their appropriateness, reliability, and contribution to the field [38]. This process ensures a deeper understanding of how data is collected, analyzed, and interpreted, providing a foundation for designing robust research approaches.

Steps to evaluate methodologies

- **Understand the methods:** Familiarize yourself with the techniques employed in the study, such as experimental setups, surveys, case studies, or computational modelling.
- **Assess appropriateness:** Determine if the chosen methodology aligns with the research objectives and adequately addresses the research questions.
- **Evaluate reliability and validity**
  1. **Reliability:** Check if the method produces consistent results across different studies or contexts.
  2. **Validity:** Ensure the method accurately measures what it intends to measure.
- **Analyze sample size and selection:** Review the sample's adequacy and representativeness. A small or biased sample can limit the generalizability of findings.

- **Review data collection techniques:** Examine the tools and processes used to gather data (e.g., surveys, interviews, or sensors). Consider whether they were suitable for capturing the necessary information.
- **Consider ethical standards:** Evaluate whether the study adhered to ethical guidelines, such as informed consent and data privacy.
- **Identify strengths and limitations:** Highlight the advantages of the approach and any shortcomings, such as potential biases, limitations in scope, or technical constraints.
- **Compare across studies:** Analyze how methodologies differ across studies and identify which approaches yielded the most reliable and insightful results.

Importance of evaluating methodologies

- **Ensures rigour:** Promotes the use of scientifically sound methods.
- **Guides research design:** Helps in selecting or improving methods for future studies.
- **Identifies gaps:** Highlights areas where methods may need refinement or innovation.

Example

In studies exploring the **impact of remote work on productivity**, different methodologies might be used:

- **Surveys:** May gather subjective employee feedback but risk bias in responses.
- **Time-tracking software:** Provides objective productivity data but could raise privacy concerns.
- **Case studies:** Offer in-depth insights but might not generalize well to other organizations.

By evaluating these methodologies, a researcher can decide whether to combine approaches, refine data collection tools, or explore entirely new methods to gain a holistic understanding of the topic.

2.3.1.3 Highlight gaps

Highlighting gaps involves identifying areas within the existing body of literature where research is insufficient, inconsistent, or outdated. Recognizing these gaps is crucial for defining novel and impactful research problems.

Steps to highlight gaps

- **Review existing studies**
    - Conduct a comprehensive review of the available literature to understand the scope and depth of research conducted so far.

- **Identify unexplored areas**
  - Look for topics or subtopics that have not been addressed or are under-researched.
- **Spot methodological limitations**
  - Note studies with outdated, flawed, or overly narrow methodologies that may require improvement or validation.
- **Analyze contradictory findings**
  - Highlight areas where different studies have produced conflicting results, indicating a need for further investigation.
- **Consider evolving trends**
  - Identify areas where new technologies, societal changes, or advancements in the field have created opportunities for updated research.
- **Examine generalizability**
  - Look for research conducted in specific contexts (e.g., limited geographical or demographic scope) that could be expanded to broader populations or settings.
- **Recognize outdated research**
  - Highlight areas where findings may no longer be relevant due to changes in technology, policy, or societal norms.

Importance of highlighting gaps

- **Guides research focus:** Directs efforts towards areas that contribute new knowledge.
- **Promotes innovation:** Encourages exploration of novel ideas and methods.
- **Increases relevance:** Ensures research aligns with current and future needs.

Example

In a review of studies on **renewable energy adoption**, gaps might include:

- Limited research on the impact of cultural attitudes on adoption rates in developing countries.
- Few studies use longitudinal data to track changes over time.
- Lack of focus on integrating renewable energy with existing infrastructure in urban areas.

By highlighting these gaps, a researcher can propose studies to address these issues, thereby advancing the field and meeting emerging needs.

2.3.1.4 Assess relationships

Assessing relationships in literature involves examining how different studies, concepts, variables, or findings interconnect. This process helps to build a comprehensive understanding of the topic, uncover patterns, and identify potential influences or dependencies.

Steps to assess relationships

- **Compare studies**
    - Identify similarities and differences in findings, methodologies, or theoretical frameworks across various studies.
- **Identify correlations**
    - Look for studies that explore relationships between key variables and assess whether they align or diverge.
- **Analyze cause-and-effect**
    - Determine if there is evidence supporting causal links or just correlations among variables in the literature.
- **Trace trends over time**
    - Examine how the relationships between concepts or findings evolve across chronological studies.
- **Map conceptual frameworks**
    - Identify shared frameworks or models used in multiple studies and how they relate to one another.
- **Evaluate interdisciplinary links**
    - Explore connections between research from different fields and their relevance to your topic.

Importance of assessing relationships

- **Contextual understanding:** Provides insight into how findings fit within the broader scope of the field.
- **Supports argumentation:** Helps justify the significance and positioning of your research.
- **Identifies new directions:** Uncovers complex relationships that might need further investigation.

Example

In reviewing the literature on AI **in healthcare**, relationships could be assessed by:

- **Comparing algorithms:** Analyzing the efficiency of different machine learning algorithms in disease prediction.
- **Identifying correlations:** Observing the link between AI implementation and patient outcomes.
- **Exploring trends:** Examining how AI applications have expanded from diagnostics to personalized medicine over time.
- **Mapping frameworks:** Connecting ethical concerns in AI use with regulations and adoption rates.

By assessing relationships, researchers can better position their study to contribute meaningfully to the existing body of knowledge and provide valuable insights.

2.3.1.5 Review outcomes

Reviewing outcomes involves synthesizing and summarizing the key insights, patterns, and conclusions derived from analyzing the literature. This step ensures that the findings from your review are clearly articulated and directly inform your research goals or objectives.

2.3.1.5.1 Steps to review outcomes

- **Summarize key findings**
  Highlight the most important results, trends, and themes identified in the literature. Focus on conclusions that align with your research question.
- **Assess strengths and weaknesses**
  Evaluate the robustness of methodologies and the reliability of conclusions in the studies reviewed. Identify well-supported findings and those that require further validation.
- **Link to research objectives**
  Connect the outcomes of your review to your research problem, ensuring they guide your methodology or hypothesis.
- **Highlight unresolved issues**
  Pinpoint gaps, contradictions, or areas where the literature lacks clarity, and propose how your research can address these.
- **Draw general conclusions**
  Provide a cohesive narrative that integrates diverse findings into a unified understanding of the topic.

Importance of reviewing outcomes

- **Guides research design:** Helps refine research questions and methodologies based on existing evidence.
- **Establishes relevance:** Demonstrates how your study builds on and contributes to existing knowledge.
- **Supports argumentation:** Provides a solid foundation for justifying the significance of your research.

Example

In a literature review on **renewable energy adoption**, the outcomes might include:

- **Key findings:** Solar and wind energy adoption has grown significantly, driven by technological advancements and policy incentives.

- **Strengths:** Studies demonstrate robust evidence of cost reduction trends.
- **Weaknesses:** Limited data on long-term environmental impacts of large-scale adoption.
- **Unresolved issues:** Insufficient research on renewable energy storage solutions.
- **General conclusion:** While adoption rates are promising, addressing storage challenges is critical for achieving global energy goals.

By systematically reviewing outcomes, researchers ensure their work is well-informed, impactful, and aligned with the existing body of knowledge.

2.3.1.6 Critique validity

Critiquing validity involves assessing the soundness, credibility, and reliability of research findings or conclusions in the literature. It ensures that the studies you rely on are methodologically robust and free from significant biases or errors. This step is crucial for establishing a strong foundation for your own research.

Aspects to consider in critiquing validity

- **Research design**
  Evaluate whether the study's design is appropriate for addressing its research question. For example, experimental studies should include proper control groups, while surveys should have representative samples.
- **Sample size and representativeness**
  Determine if the sample size is adequate and whether it accurately represents the population being studied. Small or biased samples can undermine the generalizability of findings.
- **Data collection methods**
  Assess whether the methods used to gather data were rigorous, consistent, and unbiased. Poorly defined measures or unreliable tools can compromise validity.
- **Statistical analysis**
  Check if the statistical methods used are appropriate for the type of data and research questions. Overgeneralization or misinterpretation of statistical results can lead to flawed conclusions.
- **Logical consistency**
  Evaluate whether the conclusions drawn logically follow from the data and analysis. Studies with leaps in logic or unsupported claims may lack validity.
- **Bias and confounding variables**
  Identify potential biases (e.g., selection bias and confirmation bias) or confounding factors that may have influenced the results.

- **Reproducibility**
  Assess whether the study's methods are described in sufficient detail to allow replication. Reproducibility is a hallmark of valid research.

Importance of critiquing validity

- **Ensures credibility:** Helps in relying only on trustworthy studies for your research.
- **Prevents errors:** Identifies potential weaknesses that could mislead your conclusions.
- **Guides research design:** Offers insights into how to improve your study's validity.

Example

In critiquing a study on **remote work productivity**, consider:

- **Design:** Did the study compare remote workers to in-office workers under similar conditions?
- **Sample:** Was the sample diverse in terms of industry, role, and experience?
- **Data collection:** Were self-reported productivity measures cross-validated with objective performance data?
- **Bias:** Did the researchers account for the potential bias of participants wanting to appear more productive?
- **Reproducibility:** Are the survey tools and methods clearly documented for replication?

By thoroughly critiquing validity, researchers can distinguish between reliable and unreliable sources, ensuring that their work builds on a solid foundation.

2.3.1.7 Synthesize findings

Synthesizing findings involves integrating insights from various sources to develop a coherent understanding of the research topic. This step is crucial for identifying patterns, relationships, and trends in the literature, which helps in constructing a solid foundation for your study.

Steps in synthesizing findings

- **Combine insights from multiple sources**
  Gather key points from various studies and merge them to create a unified perspective. Look for overlapping conclusions, contradictions, and complementary information.
- **Identify common themes**
  Categorize findings under major themes or subtopics that align with your research objectives. This helps in organizing the information logically.

- **Resolve conflicts in literature**
  Analyze discrepancies in the findings of different studies and attempt to explain them. Consider differences in methodologies, sample populations, or contexts.
- **Highlight patterns and trends**
  Emphasize recurring observations, such as the prevalence of a specific outcome or a consistent relationship between variables across studies.
- **Bridge gaps in knowledge**
  Use the synthesis to pinpoint gaps or unanswered questions in the literature. This sets the stage for defining your research problem.
- **Draw preliminary conclusions**
  Based on the synthesis, outline initial conclusions that provide context and justification for your study.

Importance of synthesizing findings

- **Contextual understanding:** Creates a comprehensive overview of the topic by connecting individual studies.
- **Foundation for research:** Guides the formulation of research questions and hypotheses.
- **Clarity:** Organizes diverse information into a structured narrative.

Example

When synthesizing findings on **renewable energy adoption**, consider:

- **Common themes:** Economic incentives, technological advancements, and public awareness are recurring factors influencing adoption.
- **Conflicting results:** Some studies show cost as a barrier, while others highlight it as a driver, depending on the region.
- **Trends:** Increased government policies supporting renewables over the last decade.
- **Gaps:** Limited research on the long-term economic impact of renewable energy in low-income countries.

By synthesizing findings effectively, you can present a balanced and insightful analysis of the literature, paving the way for meaningful and impactful research.

2.4 IMPORTANCE OF LITERATURE ANALYSIS

Literature analysis is a cornerstone of any research endeavour, serving multiple critical purposes that shape the research process. The flow diagram visually represents the five key roles of literature analysis, as shown in Figure 2.1. It begins with informs research design, showing how literature aids in planning and structuring the research approach. The next stage, establishes

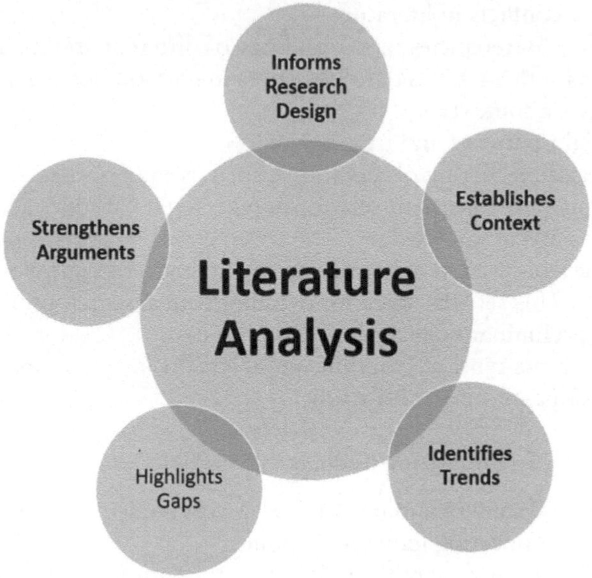

Figure 2.1 Key aspects of literature analysis in research.

context, demonstrates how literature provides background knowledge and connects the research to existing work [39]. The process continues with identifies trends, showcasing how literature highlights new directions or themes in the field. The diagram then moves to strengthens arguments, emphasizing how literature supports and validates research claims. Finally, it concludes with highlights gaps, focusing on identifying opportunities for novel contributions.

The diagram's logical flow and interconnected stages reinforce the centrality of literature analysis in producing impactful and methodologically sound research.

Its significance can be summarized as follows.

2.4.1 Informs research design

Literature analysis plays a critical role in shaping and refining the research design. It ensures that the study is built on a robust foundation of existing knowledge while identifying areas for innovation and improvement [40]. Here's how literature analysis informs research design:

1. **Defining the research problem**
   - **Role:** Literature analysis helps identify gaps, inconsistencies, or unexplored areas in previous studies, which form the basis for defining a meaningful and relevant research problem.

- **Example:** If prior research on urban air pollution focuses primarily on industrial contributions, a gap may be found in studies exploring the role of transportation policies.

2. **Refining research objectives**
   - **Role:** Analyzing the literature provides clarity on the scope and focus of the study, ensuring that objectives align with existing knowledge and address critical issues.
   - **Example:** If previous studies highlight the health impact of air pollution but lack solutions, the objective could focus on evaluating mitigation strategies.
3. **Choosing methodologies**
   - **Role:** Literature analysis provides insights into the methodologies used in similar studies, enabling researchers to adopt best practices or modify approaches for better outcomes.
   - **Example:** Studies on public health surveys might reveal the effectiveness of mixed methods (quantitative and qualitative), guiding methodology selection.
4. **Identifying variables**
   - **Role:** Helps pinpoint key variables and their relationships based on existing research, ensuring comprehensive and focused data collection.
   - **Example:** In a study on climate change awareness, literature might identify "education level" and "media exposure" as significant influencing variables.
5. **Ensuring feasibility**
   - **Role:** By reviewing the challenges and limitations faced in previous studies, literature analysis aids in designing a realistic and achievable research plan.
   - **Example:** A study requiring advanced technology might be reconsidered if literature reveals significant logistical barriers.
6. **Avoiding redundancy**
   - **Role:** Literature analysis ensures the research design does not replicate existing studies unnecessarily, promoting originality and innovation.
   - **Example:** If similar research already exists, the study could focus on a different demographic or time frame to add new insights.
7. **Justifying the research**
   - **Role:** A well-analyzed literature review provides the rationale for the study by showcasing its significance and contribution to the field.
   - **Example:** Highlighting a lack of long-term studies in renewable energy adoption justifies a longitudinal approach.

2.4.2 Establishes context

Literature analysis is vital in establishing the context for a research study, providing a solid foundation that situates the research within its broader academic or practical landscape. Here's how it helps:

1. Understanding the background
    - **Role:** It gives a detailed understanding of the history, developments, and trends related to the research topic.
    - **Example:** A study on AI in healthcare might explore its evolution from diagnostic tools to predictive analytics.
2. Defining terminology and concepts
    - **Role:** Literature analysis clarifies key terms, theories, and frameworks, ensuring consistency and precision in the study.
    - **Example:** In a study on climate change, it distinguishes between concepts like "global warming" and "carbon footprint."
3. Identifying influential works
    - **Role:** Recognizes seminal studies and influential researchers that shape the field, helping to frame the study within a lineage of knowledge.
    - **Example:** Research on human motivation might reference Maslow's hierarchy of needs as a foundational theory.
4. Positioning the study
    - **Role:** Literature analysis highlights how the research fits within the existing body of knowledge, addressing gaps or extending findings.
    - **Example:** A study on remote work could position itself by addressing mental health impacts, a topic underexplored in existing research.
5. Highlighting relevance
    - **Role:** Demonstrates the importance of the research topic by showcasing its implications and connections to real-world problems or academic debates.
    - **Example:** In cybersecurity, literature might show how data breaches affect not just individuals but also global economies.
6. Providing theoretical frameworks
    - **Role:** Literature analysis helps establish frameworks that guide the study's methodology and interpretation of results.
    - **Example:** Research on consumer behaviour could adopt the Theory of Planned Behaviour (TPB) to analyze decision-making processes.

2.4.3 Identifies trends

Literature analysis plays a crucial role in identifying trends within a research field. It reveals patterns, shifts, and emerging areas of focus that can shape and refine a researcher's approach. Here's how it contributes:

1. **Spotting emerging topics**
   - **Role:** Highlights areas gaining attention or undergoing rapid development.
   - **Example:** In renewable energy, literature might show increasing interest in green hydrogen as a sustainable alternative to fossil fuels.
2. **Tracking historical progression**
   - **Role:** Maps the evolution of a topic over time, showcasing how perspectives or technologies have changed.
   - **Example:** Studies on AI reveal a shift from rule-based systems to machine learning and deep learning over the decades.
3. **Identifying popular methodologies**
   - **Role:** Highlights commonly used methods or approaches within the field, indicating preferred practices or standards.
   - **Example:** Research in psychology may show a trend towards using mixed methods for a more comprehensive analysis.
4. **Recognizing gaps in research**
   - **Role:** Trends help pinpoint areas that remain underexplored or require further investigation.
   - **Example:** Literature on urban sustainability might reveal limited studies on smart cities in developing nations.
5. **Understanding shifts in priorities**
   - **Role:** Identifies changing priorities influenced by societal, technological, or environmental factors.
   - **Example:** The COVID-19 pandemic caused a surge in research on remote healthcare and telemedicine technologies.
6. **Guiding future research directions**
   - **Role:** Provides insights into potential future trends, enabling researchers to align their work with anticipated needs.
   - **Example:** Studies on blockchain show a growing trend towards its application in supply chain management and digital identity.

2.4.4 Strengthens arguments

Literature analysis strengthens research arguments by providing a robust foundation of evidence, context, and credibility. Here's how it enhances the quality of arguments:

1. **Provides supporting evidence**
   - **Role:** Incorporates established findings to validate claims.
   - **Example:** A study on climate change can reference prior research on rising global temperatures to support its assertions about environmental impact.
2. **Establishes credibility**
   - **Role:** Demonstrates that arguments are based on thorough research and understanding of the field.

- **Example:** Referencing peer-reviewed articles enhances the trustworthiness of claims in a thesis on public health.

3. **Connects with established theories**
   - **Role:** Links arguments to well-known theories, adding depth and coherence.
   - **Example:** A business study citing Porter's Five Forces framework strengthens its analysis of market competition.
4. **Anticipates counterarguments**
   - **Role:** Addresses opposing viewpoints by referencing literature that debates or disproves them.
   - **Example:** A paper advocating renewable energy can counter critiques by citing studies on the declining costs of solar and wind technologies.
5. **Highlights significance**
   - **Role:** Demonstrates how arguments align with or build upon important issues in the field.
   - **Example:** In education, referencing studies on digital tools emphasizes the relevance of research on e-learning solutions.
6. **Identifies trends and gaps**
   - **Role:** Shows how the research contributes to filling knowledge gaps or aligns with emerging trends.
   - **Example:** An argument for AI ethics in healthcare is strengthened by highlighting the scarcity of literature addressing this issue.
7. **Promotes logical flow**
   - **Role:** Ensures arguments are coherent and logically connected to prior research.
   - **Example:** Citing longitudinal studies to support claims about long-term effects adds logical continuity.

2.5 PLAGIARISM IN RESEARCH

Plagiarism in research refers to the unethical practice of using someone else's ideas, words, or work without proper acknowledgement. It undermines academic integrity, violates copyright laws, and damages the credibility of the researcher and their work [41]. Plagiarism can occur intentionally or unintentionally, but both forms are equally unacceptable in the research community.

2.5.1 Types of plagiarism

The flow diagram, as shown in Figure 2.2, presents a structured overview of the four primary types of plagiarism, helping to differentiate their characteristics and implications in research [41].

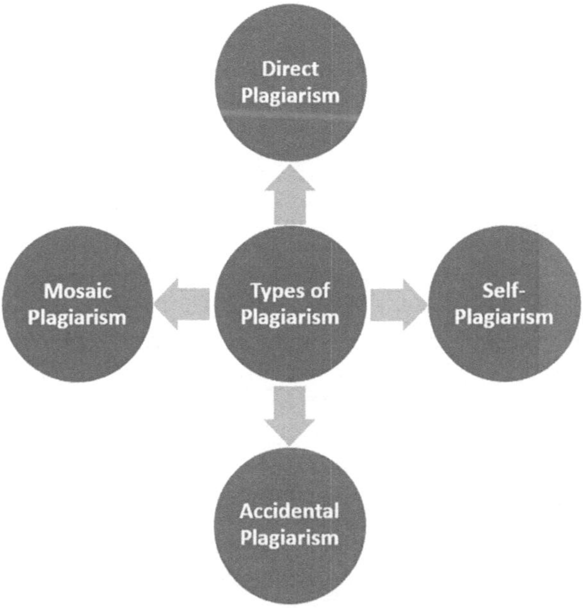

Figure 2.2 Types of plagiarism.

2.5.1.1 Direct plagiarism

Direct plagiarism is the act of copying another author's work, word-for-word, without providing proper credit or acknowledgement [42]. This unethical practice involves presenting someone else's intellectual property as your own, violating academic and professional integrity standards.

Characteristics

- No quotation marks or citations are used to indicate the original source.
- Entire sentences, paragraphs, or sections are duplicated without modification.
- Often intentional, as the individual knowingly disregards proper attribution.

Example

Original text:
"Effective communication is the cornerstone of successful teamwork."

Plagiarized text:
"Effective communication is the cornerstone of successful teamwork." (used without citation or acknowledgment).

Consequences

- Academic penalties, such as failing grades or expulsion.
- Legal repercussions if copyrighted material is involved.
- Damage to reputation and credibility in professional or academic fields.

Prevention tips

- Always cite the original source when using someone else's work.
- Use quotation marks for direct quotes.
- Employ plagiarism detection tools to verify originality before submission.

2.5.1.2 *Self-plagiarism*

Self-plagiarism occurs when an individual reuses their own previously published work or submitted content without proper acknowledgement or permission [43]. This involves presenting old work as new or original, misleading readers or evaluators.

Characteristics

- Republishing an identical or slightly modified paper in a different journal without disclosure.
- Reusing substantial parts of prior work (e.g., text, data, or figures) in a new publication without citation.
- Submitting the same assignment or project to multiple courses (known as "academic recycling").

Example

Using a paragraph from a previously published article or submitted thesis in a new research paper without citing the original work.

Consequences

- Breach of copyright agreements with publishers, potentially leading to legal issues.
- Loss of trust among peers, readers, or evaluators.
- Rejection of submissions or penalties from academic institutions.

Prevention tips

- Always disclose prior publications to editors, reviewers, or evaluators.
- Cite your earlier work when reusing any part of it in a new context.
- Avoid reusing entire sections of previous work; instead, summarize and reference them.

Self-plagiarism undermines the principle of originality in research and scholarship. While using one's own work may seem harmless, it can still violate ethical guidelines, especially when the audience expects new and innovative contributions.

2.5.1.3 Mosaic plagiarism

Mosaic plagiarism occurs when a writer borrows phrases, ideas, or sentences from a source and integrates them into their work without proper citation. Unlike direct plagiarism, the text may be rephrased or paraphrased, but the underlying ideas or structure are taken from the original source without acknowledgement.

Characteristics

- Combining words, phrases, or ideas from multiple sources into a single document without proper attribution.
- Slightly altering the original text's wording while retaining its essence.
- Misleading the reader into believing the work is entirely original.

Example

A researcher reads a study and paraphrases entire sections into their paper, altering some words but maintaining the same sentence structure and meaning, without citing the original source.

Consequences

- Breach of academic integrity policies, potentially leading to penalties.
- Damaged reputation in academic or professional circles.
- Possible retraction of published work.

Prevention tips

- Use quotation marks for direct quotes and provide appropriate citations.
- Ensure proper paraphrasing by completely rephrasing the content in your own words and citing the source.
- Keep track of all sources during the research and writing process to ensure accurate referencing.

Mosaic plagiarism can be subtle and often unintentional, but it still constitutes an ethical violation. Proper citation and a thorough understanding of the source material are essential to avoid this form of plagiarism.

2.5.1.4 Accidental plagiarism

Accidental plagiarism occurs when a researcher unintentionally fails to properly credit a source due to a lack of understanding, oversight, or errors in citation [44]. Despite the lack of intent, it is still considered a breach of academic integrity.

Characteristics

- Forgetting to cite a source used in the research.
- Improper or incomplete citation formatting.
- Paraphrasing content too closely to the original text without proper acknowledgement.
- Reusing ideas or concepts from a source without realizing the need for citation.

Example

A student writes a paper and uses information from an article but forgets to include a citation, believing it to be common knowledge, when in fact, it is not.

Consequences

- The researcher may face accusations of plagiarism, even if unintentional.
- It can harm the credibility and integrity of the work.
- Accidental plagiarism can result in rejections from journals or academic institutions.

Prevention tips

1. **Thorough note-taking:** Clearly distinguish between your own ideas and those derived from sources during the research phase.
2. **Use citation tools:** Utilize citation management software to ensure accurate referencing.
3. **Understand citation rules:** Familiarize yourself with the required citation style (e.g., APA, MLA, and Chicago).
4. **Proofread carefully:** Review your work to verify all borrowed ideas and text are properly credited.
5. **Check for plagiarism:** Use plagiarism detection software to identify unintentional overlaps with existing sources.

Accidental plagiarism highlights the importance of diligence and attention to detail in academic and professional writing. Proper citation practices and proactive checks can significantly reduce the risk of this issue.

2.5.2 Consequences of plagiarism

2.5.2.1 Reputation damage

Reputation damage refers to the loss of credibility, trust, and respect that an individual or organization experiences when accused or found guilty of plagiarism [44]. It affects personal, academic, and professional standing, often with long-term consequences.

Characteristics

- **Loss of trust:** Stakeholders, peers, or collaborators may no longer trust the individual's or organization's work.
- **Widespread impact:** Damage extends beyond the immediate context to future academic or professional relationships.
- **Public scrutiny:** Cases of plagiarism often attract negative attention, especially in academic, media, or professional circles.
- **Long-term consequences:** Even after rectifying the plagiarism, the stigma may persist, hindering future opportunities.
- **Cascading effects:** Reputational harm may affect associated parties, such as institutions, collaborators, or advisors.

Example

A professor is accused of copying significant portions of another researcher's work without citation in a published paper. As a result:

- Their university launches an investigation.
- They lose opportunities for grants and conferences.
- Students and colleagues question their integrity.
- The scandal becomes public, further tarnishing their career.

Consequences

- **Career setbacks:** Loss of job, tenure, or promotion opportunities.
- **Social ostracization:** Being excluded from academic, social, or professional networks.
- **Loss of opportunities:** Reduced chances for collaboration, funding, or leadership roles.
- **Institutional impact:** Organizations may suffer collateral damage, such as reduced credibility and trust.
- **Public backlash:** Exposure in media or academic forums can amplify the negative impact.

Prevention tips

- **Understand and follow ethical standards:** Be well-versed in ethical practices for research and writing.
- **Proper citation practices:** Attribute all sources accurately and thoroughly.
- **Use Plagiarism Detection Tools:** Regularly check work using tools like Turnitin or Grammarly.
- **Avoid over-reliance on others' work:** Strive to produce original content with proper analysis and synthesis.

- **Continuous education:** Attend workshops or seminars on research ethics and plagiarism prevention.
- **Transparent collaboration:** Maintain clear communication and documentation in collaborative efforts.

By recognizing and addressing the causes of plagiarism, individuals can avoid reputation damage and maintain credibility and trust in their academic and professional endeavours.

2.5.2.2 Legal issues

Legal issues refer to the potential lawsuits, penalties, and other legal actions that arise when plagiarism infringes upon copyright laws, intellectual property rights, or contractual agreements [45]. These issues can result in significant financial, professional, and legal repercussions for individuals or organizations.

Characteristics

- **Copyright violation:** Using someone else's work without proper authorization or attribution breaches copyright laws.
- **Contractual breach:** Failure to comply with agreements, such as publishing contracts, can lead to legal disputes.
- **Intellectual property infringement:** Unauthorized use of protected ideas, designs, or texts can result in legal action.
- **Monetary penalties:** Fines or compensations may be imposed for damages caused by plagiarism.
- **Litigation complexity:** Legal disputes may involve lengthy and costly court proceedings.

Example

A bestselling author copies substantial content from another writer's unpublished manuscript. The original author files a lawsuit for copyright infringement, seeking damages and an injunction to halt further distribution of the plagiarized book.

Consequences

- **Financial penalties:** Fines, compensation for damages, or legal fees can be substantial.
- **Loss of intellectual property rights:** The plagiarized work may be confiscated or removed from publication.
- **Injunctions:** Courts may prohibit the distribution, sale, or use of plagiarized content.

- **Criminal charges:** In severe cases, plagiarism can lead to criminal prosecution.
- **Professional sanctions:** Legal violations may result in job termination, disqualification, or blacklisting.

Prevention tips

- **Respect copyright laws:** Understand and comply with copyright and intellectual property regulations.
- **Obtain Permissions:** Secure proper permissions for the use of copyrighted material when necessary.
- **Attribute appropriately:** Always provide accurate citations and credit original authors.
- **Consult legal experts:** Seek advice from legal professionals to avoid unintentional violations.
- **Educate yourself:** Stay updated on laws related to intellectual property and plagiarism in your jurisdiction.
- **Use plagiarism detection tools:** Ensure that all work is original and properly cited before submission or publication.

By adhering to ethical and legal standards, individuals can prevent plagiarism-related legal issues, safeguard their reputations, and maintain professional integrity.

2.5.2.3 Academic penalties

Academic penalties refer to the disciplinary actions imposed by educational institutions when students, researchers, or faculty members engage in plagiarism. These penalties aim to uphold academic integrity and discourage unethical practices in academic work.

Characteristics

- **Grade reduction:** Immediate deduction in marks or failure in assignments, projects, or courses.
- **Disciplinary actions:** Formal warnings, suspension, or expulsion from the institution.
- **Loss of academic recognition:** Revocation of degrees, awards, or scholarships.
- **Damage to academic record:** Permanent notations of misconduct on transcripts.
- **Restriction on opportunities:** Disqualification from future academic programs, funding, or research roles.

Example

A student copies large sections of a published article into their thesis without proper citation. Upon detection, the institution revokes the student's degree and marks their academic record with misconduct.

Consequences

- **Failure in course:** Students may fail an assignment, project, or entire course.
- **Suspension or expulsion:** Severe cases can lead to temporary or permanent removal from the institution.
- **Revocation of credentials:** Degrees, honours, or published works may be rescinded.
- **Hindered academic progress:** Opportunities for scholarships, grants, or admissions may be lost.
- **Damaged reputation:** Academic dishonesty affects trust and credibility within the institution and beyond.

Prevention tips

- **Understand academic integrity policies:** Familiarize yourself with your institution's guidelines on plagiarism.
- **Cite properly:** Use the appropriate citation style consistently to credit all sources.
- **Original work:** Develop your own ideas and avoid over-reliance on external content.
- **Seek guidance:** Consult professors or mentors when in doubt about citation or referencing.
- **Use plagiarism checkers:** Employ tools to ensure the originality of your submissions.
- **Time management:** Plan assignments to avoid last-minute pressures that may lead to shortcuts.

By promoting ethical practices and adhering to institutional guidelines, individuals can avoid academic penalties, preserve their educational journey, and build a foundation of integrity for future success.

2.5.2.4 Hindrance to scientific progress

Hindrance to scientific progress refers to the obstruction of innovation, knowledge development, and credibility in scientific research caused by unethical practices like plagiarism. It undermines the trust in research outputs and the collaborative nature of scientific advancement.

Characteristics

- **Loss of credibility:** Research with plagiarized content is deemed unreliable and invalid.

- **Stagnation of innovation:** Duplication of ideas discourages original thinking and creativity.
- **Erosion of trust:** Repeated plagiarism damages the reputation of the research community and institutions.
- **Resource misallocation:** Time, funding, and effort are wasted on verifying or retracting flawed research.
- **Barrier to collaboration:** Plagiarism discourages partnerships and open exchange of ideas.

Example

A researcher submits a paper with plagiarized data, which gets published in a reputed journal. Upon discovery, the paper is retracted, damaging the credibility of the researcher and the institution, while also delaying advancements in that field.

Consequences

- **Retracted publications:** Journals may remove plagiarized articles, reducing the body of available knowledge.
- **Slowed innovation:** Resources spent on resolving plagiarism issues delay groundbreaking discoveries.
- **Distrust in research:** Plagiarism creates scepticism about the authenticity of scientific findings.
- **Missed opportunities:** Collaborative efforts and funding may be withheld from researchers or institutions involved in plagiarism.
- **Ethical deterioration:** The scientific community's ethical standards are weakened, discouraging younger researchers.

Prevention tips

- **Encourage originality:** Promote independent research and innovative problem-solving approaches.
- **Stringent peer review:** Implement robust peer-review processes to identify potential plagiarism early.
- **Ethics training:** Conduct regular workshops on research ethics and integrity for researchers.
- **Utilize detection tools:** Use advanced plagiarism detection software for manuscripts and proposals.
- **Acknowledge contributions:** Properly credit collaborators and previous research in citations and acknowledgements.

By addressing plagiarism and fostering a culture of integrity, the scientific community can prevent hindrances to progress, ensuring that research continues to be a reliable and innovative force for global advancement.

2.5.3 Preventing plagiarism

2.5.3.1 Proper citation

Proper citation is the ethical practice of giving credit to the original authors of ideas, data, or works used in your research [46]. It acts as a safeguard against plagiarism by ensuring transparency in the use of intellectual property.

Role in preventing plagiarism

- **Acknowledging original ideas:** Proper citation recognizes the contributions of others, making it clear which ideas are borrowed.
- **Avoiding intellectual theft:** It eliminates any ambiguity about the origin of information, ensuring no unintentional appropriation.
- **Promoting accountability:** By citing sources, researchers demonstrate accountability for their research practices.
- **Encouraging ethical practices:** Proper citation reinforces academic and professional standards of honesty.
- **Supporting verification:** It allows readers to trace and verify the authenticity of the referenced material.

Key aspects of citation to prevent plagiarism

- **Quotations:** Use quotation marks and cite the source when directly quoting text.
- **Paraphrasing:** Even when rephrasing ideas, credit the original source.
- **Summarizing:** Acknowledge the author when summarizing a work's main ideas.
- **Data and figures:** Provide citations for any data, tables, or figures borrowed from another source.
- **Multiple sources:** If multiple works influence your ideas, cite each relevant source.

Characteristics of proper citation in preventing plagiarism

- **Accuracy:** All citations should be precise, including the correct author, title, and publication details.
- **Clarity:** Clearly indicate which parts of the work rely on external sources.
- **Consistency:** Follow a standard citation format throughout the document.
- **Exhaustiveness:** Cite all materials consulted, even if not directly quoted.
- **Integrity:** Maintain ethical responsibility in referencing sources without misrepresentation.

Example

If using an idea from a journal article, a proper citation might look like:

"Renewable energy adoption has grown significantly over the past decade (Smith, 2022)."

Consequences of failing to cite properly

- **Accusations of plagiarism:** Leads to disciplinary actions or damage to reputation.
- **Loss of academic integrity:** Undermines the credibility of the research.
- **Legal implications:** In some cases, plagiarism may lead to lawsuits or copyright violations.
- **Missed academic opportunities:** Results in retractions, rejection of work, or diminished career prospects.

Tips for using proper citation to prevent plagiarism

- **Understand citation rules:** Learn the specific citation guidelines for your field (e.g., APA, MLA, and IEEE).
- **Use citation management tools:** Tools like Mendeley or EndNote can help organize and generate citations accurately.
- **Paraphrase thoughtfully:** Rewrite ideas in your own words and still credit the original author.
- **Double-check citations:** Ensure all referenced materials are correctly cited before submission.
- **Educate yourself on plagiarism:** Understand what constitutes plagiarism to avoid accidental errors.

By consistently applying proper citation practices, researchers can effectively prevent plagiarism while maintaining the integrity and credibility of their work. This approach fosters trust, upholds ethical standards, and contributes to the advancement of knowledge.

2.5.3.2 Paraphrasing skills

Paraphrasing is the process of restating a piece of text or information in your own words while retaining the original meaning [46]. When done correctly, paraphrasing helps avoid plagiarism by creating a unique version of the original content.

Role in preventing plagiarism

The primary role of paraphrasing in preventing plagiarism is to ensure that the ideas and information from external sources are presented in a new and original form. Paraphrasing allows you to integrate external knowledge into your work while giving proper credit to the original source, thus maintaining academic integrity.

Key aspects of effective paraphrasing

- **Understanding the source material:** You must fully understand the original content before attempting to paraphrase. This ensures that the meaning remains intact.
- **Rewriting in Your own words:** Simply replacing words with synonyms isn't enough. You need to change the structure and flow of the sentence while preserving the original meaning.
- **Incorporating unique expression:** You should express the idea in a way that reflects your voice or writing style, which differentiates it from the source.
- **Citation:** Even though the wording is changed, you must always provide proper citation to the original source to acknowledge where the idea came from.

Characteristics of effective paraphrasing

- **Accuracy:** The paraphrased version must accurately convey the original meaning without distorting or misrepresenting the content.
- **Originality:** The paraphrase should not resemble the original text too closely. It must be sufficiently different in structure and wording.
- **Clarity:** The paraphrased text should be clear and easy to understand, just like the original.
- **Contextual integrity:** Ensure that the paraphrase retains the context and significance of the original information.

Example

- **Original text:** "The rapid advancement of technology has significantly altered the way businesses operate, making it essential for companies to adapt quickly to stay competitive."
- **Paraphrased text:** "Technological progress has dramatically changed business operations, and for companies to remain competitive, they must quickly adjust to these changes."
- **Note:** The paraphrased text keeps the original meaning but uses different words and sentence structure.

Tips for using proper citation

- **Know the citation style:** Be familiar with the citation style required for your work (e.g., APA, MLA, and Chicago) and ensure you cite sources correctly.
- **Include in-text citations:** After paraphrasing, provide in-text citations to give credit to the original author.

- **Use quotation marks for direct quotes:** If you use the exact wording from a source, enclose it in quotation marks and provide a citation.
- **Paraphrase, don't just rearrange:** Don't simply rearrange words or phrases. Truly rewrite the content in your own voice.
- **Cite all sources:** Even if you paraphrase, you must cite all sources from which you draw information, whether directly quoted or paraphrased.

By mastering paraphrasing and citing properly, you can effectively prevent plagiarism while maintaining academic and intellectual honesty.

2.5.3.3 Plagiarism detection tools

Plagiarism detection tools are software or online services designed to help identify instances of copied or unoriginal content in academic, professional, or creative works [47]. These tools compare the text to various online sources, databases, and publications to check for similarities and potential plagiarism.

Role of plagiarism detection tools

- **Identify plagiarism:** These tools highlight passages that are similar or identical to existing sources, helping to detect both direct copying and improper paraphrasing.
- **Ensure originality:** They help writers, researchers, and students ensure that their work is original and that all sources are appropriately cited.
- **Prevent academic dishonesty:** By checking for plagiarism, these tools help maintain integrity in academic and professional writing.

Key features of plagiarism detection tools

- **Text comparison:** They compare the submitted text against a vast database of sources, including academic journals, websites, books, and more.
- **Percentage of similarity:** Most tools provide a percentage indicating how much of the text matches existing sources.
- **Source identification:** They often identify the sources of the matched content, allowing users to trace the origin of the similarities.
- **Reports:** These tools generate detailed reports that highlight the plagiarized sections and show the corresponding sources.

Popular plagiarism detection tools

- **Turnitin**
  - **Overview:** One of the most widely used plagiarism detection tools in academic institutions.
  - **Key features:** Compares submitted papers against a vast database of academic content, student papers, and publications. It also provides similarity reports and originality checks.

- Use: Primarily used by universities and educational institutions.
- **Grammarly**
  - **Overview:** A comprehensive writing assistant tool with plagiarism detection features.
  - **Key features:** Checks for both grammatical errors and plagiarism by comparing the text against billions of web pages.
  - **Use:** Ideal for both casual and professional writing.
- **Copyscape**
  - **Overview:** A popular tool for checking plagiarism in web content.
  - **Key features:** Detects instances of content duplication on the internet, often used by bloggers, content creators, and marketers.
  - **Use:** Best suited for checking website content for originality.
- **Plagscan**
  - **Overview:** A plagiarism detection tool used by both educational institutions and businesses.
  - **Key features:** Provides detailed plagiarism reports and matches text against academic journals, websites, and other sources.
  - **Use:** Often used by researchers and companies to check academic and professional content.
- **Quetext**
  - **Overview:** A user-friendly plagiarism detection tool that offers both free and paid versions.
  - **Key features:** It uses DeepSearch technology to check for plagiarism and offers a citation assistant to help users with proper attribution.
  - **Use:** Suitable for both students and professionals.
- **DupliChecker**
  - **Overview:** A free plagiarism detection tool with a simple interface.
  - **Key features:** Provides free daily checks with a limited word count and paid options for larger documents.
  - **Use:** Useful for casual users or those who need to check shorter texts.
- **Plagiarism checker (small SEO tools)**
  - **Overview:** A free plagiarism checker primarily used for SEO purposes.
  - **Key features:** Checks for duplicate content across the web and offers a detailed report.
  - **Use:** Popular with content creators and SEO specialists.

Characteristics of plagiarism detection tools

- **Accuracy:** The ability to accurately identify copied or closely paraphrased content.
- **Comprehensiveness:** The scope of the tool's database, including academic papers, web pages, and publications.

- **Ease of use:** User-friendly interfaces that make it easy for individuals to upload documents and receive reports.
- **Cost:** Some tools offer free versions with limitations, while others require a subscription or one-time payment for full access.
- **Integration:** Many tools integrate with word processors or learning management systems (LMS) for seamless plagiarism checks.

Example

A student submits an essay to Turnitin, and the tool detects a 20% similarity with a published article from an academic journal. The student receives a detailed report showing the matching sections and their sources. The student can then revise the essay to paraphrase the content correctly and cite the source properly, ensuring academic integrity.

Tips for using plagiarism detection tools

- **Check early:** Run your work through a plagiarism checker early in the writing process to identify potential issues before final submission.
- **Paraphrase properly:** If a tool highlights similarities, make sure to paraphrase properly and cite the source.
- **Don't rely solely on tools:** Plagiarism checkers are useful, but they can't always detect all forms of plagiarism, such as self-plagiarism or improperly paraphrased content.
- **Review the report:** Always read the plagiarism report carefully and manually verify the highlighted sections to ensure proper citation and paraphrasing.
- **Use multiple tools:** Sometimes using more than one plagiarism detection tool can provide a more comprehensive check.

By using plagiarism detection tools effectively, you can maintain originality, avoid unintentional plagiarism, and uphold academic and professional integrity.

2.5.3.4 Maintaining research integrity

Research integrity refers to the adherence to ethical principles and professional standards in the conduct of research. It ensures that the research process is transparent, accurate, and responsible and that the findings are credible and trustworthy.

Key aspects of research integrity

- **Honesty:** Researchers should be truthful about their data, methods, and findings. Fabricating or falsifying results is a severe violation of research integrity.

- **Accuracy**: All aspects of the research, from data collection to analysis, should be performed with the utmost precision. Misrepresentation of data or results can lead to incorrect conclusions.
- **Transparency**: Researchers must be open about their methods, data, and funding sources. Transparency ensures that others can replicate the research and verify the results.
- **Accountability**: Researchers are accountable for their work, ensuring it is conducted ethically, responsibly, and in accordance with legal and professional standards.
- **Fairness**: Proper acknowledgement and credit should be given to the contributions of others, and conflicts of interest should be disclosed.
- **Respect for intellectual property**: Researchers must respect intellectual property rights, including giving proper credit for ideas, theories, and work of others through citations.

Characteristics of research integrity

- **Ethical decision-making**: Researchers should follow ethical guidelines in their work, including respecting human participants, ensuring privacy, and maintaining objectivity.
- **Peer review and validation**: Submitting research for peer review helps ensure the work meets high standards and is critically assessed by experts in the field.
- **Proper documentation**: Researchers should document their methodology, data collection, and analysis procedures clearly and in detail to allow for reproducibility.
- **Avoiding bias**: Efforts should be made to eliminate personal, financial, or institutional biases from the research process, ensuring that results are objective and impartial.

Examples of violations of research integrity

- **Plagiarism**: Copying or paraphrasing another researcher's work without proper attribution.
- **Data fabrication/falsification**: Creating false data or altering data to support a desired conclusion.
- **Selective reporting**: Omitting or ignoring data that contradicts the hypothesis or expected outcome.
- **Conflict of interest**: Failing to disclose financial or personal interests that may influence the research process or interpretation of results.

Steps to maintain research integrity

- **Follow ethical guidelines**: Always adhere to established ethical standards and guidelines, such as those set by Institutional Review Boards (IRBs) or professional associations.

- **Properly cite sources:** Ensure all sources of information, ideas, and data are properly credited. This includes paraphrased material, direct quotes, and any contributions from colleagues.
- **Be transparent about methodology:** Provide a clear, detailed description of your research methodology so others can replicate your work and validate your findings.
- **Use reliable data:** Ensure the data you use is accurate, relevant, and from credible sources. Avoid manipulating or selectively presenting data to fit a predetermined narrative.
- **Submit for peer review:** Submit your work to peer-reviewed journals or conferences to get feedback from experts, ensuring the quality and integrity of your research.
- **Maintain confidentiality:** When conducting research involving sensitive data, such as human subjects or proprietary information, ensure confidentiality and respect for privacy.
- **Disclose conflicts of interest:** Be transparent about any financial, professional, or personal relationships that could be perceived as a conflict of interest in your research.
- **Proper data management:** Maintain accurate records of your data and methodology, including backups, so that your research can be verified and reproduced by others.

Importance of research integrity

- **Trust in research:** Research integrity ensures that the scientific community, policymakers, and the public can trust the results and conclusions drawn from research.
- **Credibility:** Maintaining high standards of research integrity strengthens the credibility of the researcher and their institution.
- **Advancement of knowledge:** Ethical research practices contribute to the accurate and honest advancement of knowledge, allowing the scientific community to build upon reliable findings.
- **Prevention of harm:** By maintaining integrity, researchers prevent the dissemination of misleading or false information, which can have harmful consequences, especially in fields like medicine, engineering, and public policy.

Tips for maintaining research integrity

- **Regularly update your knowledge:** Stay informed about ethical guidelines, best practices, and new developments in your field.
- **Work collaboratively:** Engage in collaborative research, but ensure that contributions are properly acknowledged and work is divided equitably.
- **Use plagiarism detection tools:** Before submitting your work, use plagiarism detection tools to ensure proper attribution and avoid unintentional plagiarism.

- **Seek mentorship:** Engage with experienced researchers who can guide you in maintaining ethical practices and help you navigate complex research issues.
- **Adopt transparent reporting:** Always disclose all results, even if they don't support your hypothesis, and provide full transparency regarding your research process.

By adhering to these principles and practices, researchers can maintain the integrity of their work and contribute positively to the academic and scientific communities.

2.5.3.5 Clear documentation

Clear documentation refers to the thorough and transparent recording of all aspects of the research process, including methodology, data collection, analysis, results, and conclusions. It ensures that research can be easily understood, verified, and replicated by others, contributing to the integrity and credibility of the work.

Importance of clear documentation

- **Reproducibility:** Clear documentation allows other researchers to replicate the study and verify the findings, which is a cornerstone of scientific progress.
- **Transparency:** It provides transparency in how research was conducted, ensuring that no steps are hidden or manipulated.
- **Accountability:** Researchers can be held accountable for their methods and decisions, which ensures the accuracy and integrity of the research.
- **Legal and ethical compliance:** Proper documentation helps ensure that the research complies with ethical standards, legal requirements, and funding agency expectations.
- **Sharing knowledge:** Well-documented research makes it easier for others to understand and build upon the findings, advancing the field of study.

Key aspects of clear documentation

- **Research objectives and hypothesis:** Clearly state the research goals, objectives, and hypotheses. This provides a roadmap for the study and guides the methodology.
- **Methodology:** Document all research methods in detail, including:
  1. **Experimental design:** Describe how the study was structured (e.g., sample size and control groups).
  2. **Data collection procedures:** Outline how data was gathered (e.g., surveys, experiments, and interviews).

3. **Data analysis:** Specify the techniques and tools used for analyzing the data (e.g., statistical methods and software tools).

- **Data management:** Maintain accurate records of raw data, including:
  1. **Data sources:** Document where the data came from, including any datasets or sources consulted.
  2. **Data storage:** Ensure data is securely stored and backed up, and describe how it can be accessed by others.
  3. **Data transformations:** If the data is processed or cleaned, document how it was manipulated.
- **Results and findings:** Present the results of your research in a clear and structured format, such as tables, graphs, or charts. Be honest about both positive and negative results.
- **Citations and references:** Provide complete citations for all sources of information, including primary literature, datasets, and other relevant materials used in your research. Follow a consistent citation style (e.g., APA, MLA, and Chicago).
- **Ethical considerations:** Document any ethical approvals or consents obtained (e.g., IRB approval, informed consent from participants).
- **Challenges and limitations:** Acknowledge any challenges or limitations faced during the research process. This could include issues with data collection, methodology, or unforeseen factors that may have influenced the results.

Characteristics of clear documentation

- **Clarity:** The documentation should be written in clear, concise language, free of jargon, so that others can easily follow the research process.
- **Completeness:** All relevant information, including raw data, methods, and analysis steps, should be included to provide a full picture of the research.
- **Organization:** The documentation should be well-organized, with sections clearly labelled and logically ordered. This makes it easier for others to navigate and understand.
- **Consistency:** Ensure that terminology, formatting, and citation style are consistent throughout the document.
- **Transparency:** Avoid omitting any details that could affect the interpretation or replication of the research.

Best practices for clear documentation

- **Maintain a research log:** Keep a daily or weekly log of research activities, including decisions made, problems encountered, and adjustments to the methodology. This helps ensure that the process is fully documented in real time.

- **Use data management tools:** Use tools like spreadsheets, databases, or data management software to keep track of data and its transformations. Label files clearly and consistently.
- **Create detailed protocols:** For experimental or fieldwork research, create step-by-step protocols for how data should be collected and analyzed, ensuring consistency throughout the study.
- **Use version control:** For large or collaborative projects, use version control systems (e.g., Git) to track changes in documents, code, or data files.
- **Share raw data and code:** When possible, share raw data and analysis code in open-access repositories, allowing others to verify your work and build upon it.
- **Review and revise:** Regularly review your documentation to ensure that it is accurate and up-to-date. Revise sections that may be unclear or incomplete.
- **Document assumptions:** Clearly state any assumptions made during the research process (e.g., assumptions about data quality or participant behaviour).

Example of clear documentation

- **Research objective:** Investigating the effects of different light intensities on plant growth.
- **Methodology**
  - Plants were grown under three different light intensities (low, medium, and high) for six weeks.
  - Growth was measured by tracking the height of the plants every week.
  - Statistical analysis was performed using ANOVA to compare growth rates across light intensities.
- **Data**
  - Raw data was collected in a spreadsheet, with columns for plant ID, light intensity, week, and height measurement.
  - Data cleaning included removing any outliers or measurements that were clearly incorrect.
- **Results**
  - The average growth rate was highest under medium light intensity, with a significant difference between the low and high intensities.
- **Ethical considerations**
  - No ethical approval was required for this study as it involved non-sentient plants.

Tips for effective documentation

- **Start early:** Begin documenting your research process as soon as possible, even during the planning stages.

- **Be consistent:** Use the same format and style for documenting data, methods, and results across all sections of your research.
- **Use clear labels:** Label all figures, tables, and charts with descriptive titles and provide clear explanations in the captions.
- **Provide access to raw data:** If possible, upload your raw data and code to open-access platforms (e.g., Figshare and GitHub) for transparency and reproducibility.
- **Use standardized templates:** For large projects, use standardized templates for data collection and analysis to ensure uniformity across the research process.

By maintaining clear documentation throughout the research process, you enhance the transparency, reliability, and reproducibility of your work, contributing to the credibility of the scientific community and ensuring that your research can be verified and built upon by others.

2.6 RESEARCH ETHICS

Research ethics refers to the moral principles and guidelines that govern the conduct of research, ensuring that studies are carried out responsibly and with respect for participants, data, and society [48]. It aims to ensure that research is done with integrity, fairness, and accountability, prioritizing the well-being of individuals and communities involved.

2.6.1 Key principles of research ethics

2.6.1.1 Integrity

In the context of **research ethics**, integrity refers to the commitment to conducting research in an honest, transparent, and responsible manner, ensuring that all actions, decisions, and findings adhere to the highest ethical standards. It is a cornerstone of research ethics because it guarantees that research is both credible and trustworthy.

Key aspects of integrity in research ethics

- **Honesty**
  - Researchers must report their findings truthfully, whether the results support or contradict the hypothesis. Misleading or fabricated data, such as falsifying results or manipulating data to fit desired outcomes, is a breach of integrity and ethics.
- **Transparency**
  - Integrity in research requires full disclosure of the research process, including methodology, data collection methods, and potential conflicts of interest. Transparent reporting allows others to verify

and replicate the study, which is fundamental for maintaining trust in scientific findings.

- **Accountability**
  - Researchers must take responsibility for their work, ensuring that their methods, data, and conclusions are justifiable and accurate. Accountability also means being open to criticism and revising work when necessary, based on feedback or new evidence.
- **Accuracy**
  - Accurate data collection, analysis, and reporting are essential to maintain integrity. Any errors or discrepancies should be addressed openly, and the correct procedures should be followed to ensure that the results are valid.
- **Fairness**
  - Fairness in research involves giving proper credit to others' contributions, acknowledging sources appropriately, and avoiding plagiarism. Researchers should also treat participants, colleagues, and competitors fairly and without bias.
- **Non-maleficence** (Do No Harm)
  - A researcher's integrity includes the ethical responsibility to minimize harm to participants, the environment, and society. Ethical research practices involve ensuring the safety and well-being of all individuals involved in the study.

Importance of integrity in research ethics

- **Credibility and trustworthiness**
  - Research conducted with integrity builds credibility. When researchers are honest and transparent, their work is more likely to be trusted by the scientific community, funding agencies, and the public.
- **Scientific progress**
  - Research integrity is essential for the accurate accumulation of knowledge. Misleading or false information can hinder progress, whereas honest research allows the scientific community to build upon reliable findings.
- **Public trust**
  - Maintaining integrity ensures that research is conducted for the benefit of society. It fosters trust in scientific research, especially in areas like medicine, environmental science, and technology, where unethical practices can have far-reaching consequences.
- **Ethical responsibility**
  - Integrity is at the heart of ethical research. Researchers have a moral obligation to uphold high standards in their work, protecting participants, respecting intellectual property, and ensuring that their research serves the greater good.

Example of integrity in research ethics

A researcher studying the effectiveness of a new drug should:

- Report all data accurately, even if the drug did not work as expected.
- Disclose any conflicts of interest, such as financial ties to the drug manufacturer.
- Ensure that the methodology is clear and replicable by others in the field.
- Obtain informed consent from all participants and maintain their privacy.
- Address any errors in data or methodology openly and revise findings when necessary.

2.6.1.2 *Confidentiality*

Confidentiality in research ethics refers to the obligation of researchers to protect sensitive information obtained during the course of a study, ensuring that personal, private, or proprietary data is not disclosed to unauthorized individuals or entities. It is a critical aspect of ethical research, safeguarding the privacy of participants and maintaining the trust of those involved in the research process.

Key aspects of confidentiality in research

- **Protection of participant information**
  - Researchers must ensure that personal information collected from participants, such as names, contact details, and responses, remains confidential. This is especially important in studies involving sensitive data, such as medical or psychological research.
- **Anonymity vs. confidentiality**
  - **Anonymity** refers to ensuring that participants' identities are not linked to the data collected, even by the researchers.
  - **Confidentiality** means that the researcher knows the identities of the participants but ensures that this information is not disclosed to others without consent.
  - While both protect participant privacy, confidentiality allows for a more nuanced and flexible approach to handling sensitive data.
- **Data security**
  - Researchers must take appropriate measures to protect the data from unauthorized access, loss, or theft. This includes using secure storage systems, encryption, and password protection for digital data and physical safeguards for paper records.
- **Informed consent**
  - During the informed consent process, researchers must explain to participants how their data will be used, who will have access to it,

and the measures taken to ensure confidentiality. Participants must consent to these terms before data collection begins.

- **Limitations of confidentiality**
    - In certain cases, confidentiality may be breached, such as when disclosure is required by law (e.g., in cases of child abuse or imminent harm to others). Researchers must inform participants about these potential exceptions during the consent process.
- **Data deletion or destruction**
    - Once the research is completed, researchers should destroy or anonymize any data that is no longer needed to ensure the continued protection of participants' privacy.

Importance of confidentiality in research ethics

- **Protecting participants' privacy**
    - Confidentiality ensures that personal information and sensitive data shared by participants are protected from exposure, safeguarding their privacy and dignity.
- **Building trust**
    - Maintaining confidentiality fosters trust between the researcher and participants. Participants are more likely to be honest and forthcoming with their information if they know their privacy will be respected.
- **Encouraging participation**
    - When participants feel their information will be kept confidential, they are more likely to agree to participate in studies, especially in sensitive research areas (e.g., mental health, sexual behaviour, or illegal activities).
- **Ethical responsibility**
    - Researchers have a duty to respect the privacy of individuals and to ensure that any information collected is handled ethically. This includes ensuring that any data shared with third parties is anonymized or aggregated to prevent identification.

Example of confidentiality in research ethics

In a clinical trial for a new drug:

- **Personal information:** The names, addresses, and health records of participants are kept confidential and only accessible to the research team.
- **Anonymization:** Data may be anonymized for analysis, with identifiers removed to prevent any linkage to specific individuals.
- **Informed consent:** Participants are informed that their data will be kept confidential and are made aware of any exceptions (e.g., reporting of adverse effects to regulatory bodies).
- **Data security:** All data, whether in paper or digital form, is stored in a secure manner, such as encrypted files or locked cabinets.

Best practices for ensuring confidentiality

- **Clear communication:** Researchers should clearly explain confidentiality policies to participants during the informed consent process.
- **Secure data handling:** Use encrypted systems, password protection, and other security measures to safeguard digital and physical data.
- **Limit access:** Only authorized personnel should have access to sensitive data, and access should be limited to the minimum necessary.
- **Data minimization:** Collect only the data necessary for the research and avoid collecting excessive or irrelevant information that could increase privacy risks.
- **Training:** Researchers and staff should be trained in confidentiality practices to ensure that all ethical guidelines are followed.

2.6.1.3 Informed consent

Informed consent is a fundamental ethical principle in research that ensures participants voluntarily agree to take part in a study, fully understanding the nature of the research, its potential risks, and their rights. It is an essential aspect of protecting participants' autonomy and ensuring ethical standards in research.

Key aspects of informed consent

- **Voluntary participation**
  - Participants must freely choose to participate without any coercion, manipulation, or undue influence. They should have the right to withdraw from the study at any time without facing any negative consequences.
- **Full disclosure**
  - Researchers must provide participants with all relevant information about the study. This includes the purpose of the research, procedures involved, potential risks, benefits, and the duration of the study. The information should be presented in clear, understandable language, free from technical jargon.
- **Understanding**
  - Researchers must ensure that participants fully understand the information provided. This may involve providing additional explanations or offering consent documents in languages or formats that are accessible to the participants.
- **Capacity to consent**
  - Participants must have the mental and legal capacity to provide informed consent. If participants are unable to give consent (e.g., minors and individuals with cognitive impairments), a legally authorized representative (e.g., parent or guardian) may provide consent on their behalf.

- **Right to withdraw**
  - Participants must be informed that they have the right to withdraw from the study at any time, without penalty or loss of benefits. Their decision to withdraw should be respected and not lead to any negative consequences.
- **Confidentiality and privacy**
  - Participants should be informed about how their personal data will be protected, who will have access to it, and how it will be used. They must also be informed about any exceptions to confidentiality, such as mandatory reporting requirements.
- **Risks and benefits**
  - Researchers should clearly explain any potential risks, discomforts, or adverse effects associated with participation in the study. At the same time, the benefits of the research should be communicated, both for the participants and the broader community or field of study.

Importance of informed consent in research ethics

- **Protects participant autonomy**
  - Informed consent respects the right of individuals to make decisions about their own participation in research. It ensures that participants are active decision-makers rather than passive subjects.
- **Promotes trust**
  - By fully informing participants and respecting their autonomy, researchers foster trust and transparency, which are essential for ethical research practices.
- **Ethical and legal requirements**
  - Informed consent is not only an ethical obligation but also a legal requirement in many research contexts. Failure to obtain informed consent can lead to ethical violations and legal consequences.
- **Minimizes harm**
  - By ensuring participants are aware of potential risks and discomforts, informed consent helps minimize harm and ensures that participants are making an informed decision about their involvement.
- **Enhances research validity**
  - Ethical research practices, including informed consent, contribute to the validity of findings by ensuring that participants willingly and knowingly contribute to the study, which reduces the likelihood of biased or unrepresentative data.

Example of informed consent in research ethics

In a clinical trial for a new drug:

- **Disclosure:** Participants are informed about the purpose of the study, the experimental nature of the drug, the potential side effects, and the duration of the trial.

- **Understanding:** Researchers explain the information in lay terms and check that participants understand the risks, benefits, and their rights.
- **Voluntary participation:** Participants are told they can withdraw from the study at any time without penalty and that their treatment will not be affected by their decision to withdraw.
- **Confidentiality:** Participants are informed that their personal information will be kept confidential and that their identities will not be disclosed in any publications or reports.
- **Signature:** Participants sign a consent form acknowledging that they have understood the information provided and voluntarily agree to participate.

Best practices for informed consent

- **Provide clear and comprehensive information:** Ensure that the consent form or process covers all essential aspects of the study, including purpose, procedures, risks, benefits, and confidentiality.
- **Use simple language:** Avoid technical jargon or complex terminology. Information should be understandable to the target population, considering their literacy levels and language skills.
- **Offer time for questions:** Participants should have the opportunity to ask questions and receive answers before agreeing to participate. Researchers should address any concerns participants may have.
- **Ensure voluntary consent:** Participation must be voluntary, and participants should not feel pressured to join or remain in the study. They should also be made aware that they can withdraw at any time.
- **Obtain consent in writing:** Document the informed consent process by having participants sign a consent form. This serves as a record that the participant has been informed and has agreed to participate.
- **Monitor ongoing consent:** For long-term or ongoing studies, periodically check that participants still consent to participate, especially if the nature of the study changes over time.

Limitations and considerations

- **Cognitive or language barriers:** In some cases, participants may not fully understand the consent process due to cognitive impairments or language differences. Researchers must find ways to accommodate these participants, such as using translators or providing additional support.
- **Vulnerable populations:** Special considerations are needed when obtaining consent from vulnerable populations (e.g., minors, individuals with mental disabilities, or those in dependent relationships), requiring additional safeguards and sometimes parental or guardian consent.

2.6.1.4 Non-maleficence

Non-maleficence is a core ethical principle in research that mandates researchers to avoid causing harm to participants. Derived from the Hippocratic oath ("do no harm"), this principle ensures that researchers prioritize the safety, well-being, and dignity of participants throughout the study.

Key aspects of non-maleficence in research ethics

- **Minimizing harm**
  - Researchers must design studies in a way that minimizes potential risks or discomforts to participants. Any potential harm—whether physical, psychological, emotional, or social—should be anticipated and minimized through careful planning and safeguards.
- **Risk assessment:**
  - Researchers must conduct a thorough risk assessment before starting the study. This includes identifying potential harms, weighing the benefits of the research against these risks, and implementing measures to reduce or eliminate harm.
- **Avoiding unnecessary risks:**
  - If a study presents risks that are not essential to answering the research question, those risks should be avoided. Researchers should seek alternative methods that pose less harm to participants when possible.
- **Ethical review and oversight:**
  - Before beginning any research, especially studies involving human participants, ethical review boards (e.g., Institutional Review Boards or IRBs) should evaluate the study's design to ensure that the risks are minimized and ethically justified.
- **Monitoring and adjusting**
  - Throughout the research process, researchers should continually monitor the well-being of participants. If unforeseen risks or harms arise during the study, researchers must take immediate steps to mitigate those harms, such as halting the study or adjusting protocols.
- **Informed consent**
  - Participants must be made aware of any potential risks involved in the study during the informed consent process. Researchers are responsible for fully disclosing risks and ensuring participants understand them before agreeing to participate.

Importance of non-maleficence in research ethics

- **Protects participants' well-being**
  - The primary goal of non-maleficence is to safeguard the health and welfare of participants. Researchers must ensure that participants

are not exposed to undue harm or discomfort, physically or emotionally.

- **Promotes ethical research**
  - Non-maleficence reinforces the ethical responsibility of researchers to prioritize participants' safety, ensuring that research is conducted with care and respect for human life and dignity.
- **Builds trust**
  - By adhering to non-maleficence, researchers help build trust with participants and the broader public. When people know that their well-being is a top priority, they are more likely to participate in research and trust the outcomes.
- **Minimizes negative consequences:**
  - Ethical research practices reduce the risk of negative consequences, such as harm to participants, damage to public perception of research, or legal repercussions for researchers and institutions.

Example of non-maleficence in research ethics

In a clinical trial for a new drug:

- **Risk assessment:** Researchers assess potential side effects of the drug and ensure that the risks to participants are minimal compared to the potential benefits.
- **Informed consent:** Participants are fully informed of any known risks (e.g., nausea and headaches) and agree to participate with the understanding of these potential side effects.
- **Monitoring:** During the trial, researchers monitor participants closely for any adverse reactions and take action if any harmful effects are observed, such as stopping the trial or providing medical intervention.
- **Safeguards:** If the drug poses any unexpected or severe risks, researchers halt the trial to prevent further harm to participants.

Best practices for ensuring non-maleficence

- **Risk–benefit analysis**
  - Before conducting research, assess whether the potential benefits outweigh the risks. If the risks are too high, the study should not proceed.
- **Informed consent process**
  - Ensure participants are fully informed about the risks involved in the study and that they have the right to withdraw at any time without facing negative consequences.
- **Continuous monitoring**
  - Throughout the research process, continually monitor participants for any adverse effects and take immediate action if harm is detected.

- **Ethical review**
  - Seek approval from an ethical review board (e.g., IRB) to ensure that the research is designed to minimize harm and is ethically justified.
- **Debriefing**
  - After the study, especially if there is any risk of psychological or emotional harm, debrief participants and offer support if needed.

2.6.1.5 Respect for participants

Respect for participants is a fundamental ethical principle in research that acknowledges the inherent dignity and autonomy of individuals. It ensures that researchers treat participants with courtesy, fairness, and consideration throughout the research process, safeguarding their rights and well-being.

Key aspects of respect for participants

- **Autonomy and voluntary participation**
  - Participants must have the freedom to make their own decisions about whether to participate in a study, without any form of coercion, manipulation, or undue influence. Researchers must respect the autonomy of participants by providing them with all necessary information to make an informed decision.
- **Informed consent**
  - As part of respecting participants, researchers must obtain informed consent before any data is collected. This means clearly explaining the study's purpose, procedures, risks, and benefits, ensuring that participants understand their involvement and can make a voluntary decision to participate.
- **Confidentiality and privacy**
  - o Respect for participants includes protecting their personal information and ensuring that any data collected is kept confidential. Researchers must guarantee that participants' privacy is respected and that their data is not shared without their consent.
- **Right to withdraw**
  - Participants should be informed that they have the right to withdraw from the study at any time without facing negative consequences. Their decision to leave the study must be respected, and their data should be handled according to their wishes.
- **Cultural sensitivity**
  - Researchers must be sensitive to the cultural, social, and individual backgrounds of participants. This involves understanding and respecting participants' values, beliefs, and preferences, especially in studies involving diverse or vulnerable populations.

- **Fair treatment**
  - Participants must be treated with fairness and equality throughout the research process. This includes avoiding discrimination and ensuring that the study's benefits and risks are distributed equitably among participants.
- **Protection from harm**
  - Respect for participants involves ensuring that they are not exposed to unnecessary risks or harm during the research. Researchers must prioritize participants' well-being and minimize any potential physical, emotional, or psychological harm.
- **Debriefing**
  - After the research is completed, participants should be provided with a debriefing, particularly if the study involved deception or if there were any risks of psychological harm. The debriefing should clarify the study's purpose, address any concerns, and provide resources if necessary.

Importance of respect for participants

- **Ethical responsibility**
  - Researchers have an ethical duty to treat participants with respect, acknowledging their autonomy, dignity, and rights. This principle helps ensure that research is conducted in a manner that upholds the moral values of fairness and justice.
- **Trust and cooperation**
  - By respecting participants, researchers build trust, which is essential for encouraging participation in future studies. Trust in the research process leads to more accurate and reliable data, as participants are more likely to be honest and forthcoming.
- **Minimizes harm and maximizes benefits**
  - When participants are treated with respect, their well-being is prioritized, which minimizes harm and ensures that the research benefits them and the broader community.
- **Legal and regulatory compliance**
  - Respect for participants ensures that researchers comply with ethical and legal standards, such as obtaining informed consent, maintaining confidentiality, and following regulations for research involving human subjects.

Example of respect for participants in research

In a psychological study on stress:

- **Informed consent:** Participants are fully informed about the nature of the study, the potential risks of stress induction, and their right to withdraw at any time without penalty.
- **Confidentiality:** All personal data is kept confidential, and participants' identities are protected throughout the study.

- **Cultural sensitivity:** The researchers ensure that the study design and procedures are culturally sensitive, taking into account the participants' backgrounds and ensuring no discomfort or distress.
- **Right to withdraw:** Participants are made aware that they can leave the study at any time, and their data will not be used if they choose to withdraw.
- **Debriefing:** After the study, participants are debriefed, given information on the purpose of the study, and provided with resources to manage any stress they may have experienced during the study.

Best practices for ensuring respect for participants

- **Clear communication**
  - Researchers should communicate openly and honestly with participants, ensuring they understand the purpose, procedures, and potential risks of the study.
- **Cultural competence**
  - Researchers should strive to understand and respect the cultural backgrounds of participants, adapting their approach as necessary to ensure that the study is conducted in a respectful and inclusive manner.
- **Confidentiality protections**
  - Researchers must implement strict measures to protect participants' personal data, ensuring that their information is kept secure and used only for the purposes of the research.
- **Ongoing consent**
  - Researchers should continuously ensure that participants remain informed and that they have the opportunity to withdraw at any time. This includes revisiting the consent process if the study changes or if new risks arise.
- **Participant support**
  - Researchers should offer support to participants throughout the study, including providing resources for psychological or emotional assistance if needed, especially in studies that may induce stress or discomfort.

2.6.1.6 Accountability

Accountability in research ethics refers to the responsibility of researchers to ensure that their work is conducted ethically, transparently, and with integrity. It involves being answerable for all aspects of the research process, from design and data collection to analysis and reporting. Researchers must take responsibility for the ethical conduct of their work, including ensuring compliance with legal and institutional regulations and addressing any concerns raised about their research practices.

Key aspects of accountability in research

- **Responsibility for ethical conduct**
  - Researchers are accountable for conducting their research in accordance with ethical guidelines, including obtaining informed consent, protecting participant confidentiality, minimizing harm, and avoiding conflicts of interest.
- **Transparency**
  - Researchers must ensure transparency in their research methods, findings, and funding sources. This includes clearly reporting research procedures, data analysis, and any limitations or biases in the study.
- **Honesty in reporting**
  - Researchers are responsible for presenting their findings truthfully, without fabrication, falsification, or misrepresentation. This includes acknowledging any limitations of the study and being open about potential conflicts of interest.
- **Compliance with regulations**
  - Researchers must adhere to institutional, local, national, and international regulations governing research, such as those set by ethical review boards (IRBs), funding bodies, and professional organizations.
- **Addressing misconduct**
  - If any research misconduct or ethical violations are identified, researchers are accountable for addressing them. This may involve investigating the issue, correcting the record, or reporting misconduct to relevant authorities.
- **Respecting participant rights**
  - Researchers are accountable for respecting participants' rights, including obtaining informed consent, protecting their privacy, and ensuring that participants are not exposed to unnecessary risks.
- **Academic integrity**
  - Researchers must maintain academic integrity by giving proper credit to the contributions of others, avoiding plagiarism, and acknowledging sources and prior work in their research.

Importance of accountability in research ethics

- **Maintaining trust**
  - Accountability is essential for maintaining trust between researchers, participants, and the public. Transparent and ethical research practices ensure that participants feel confident that their rights are being respected and that the research is being conducted in a trustworthy manner.

- **Ensuring research integrity**
  - Researchers who are accountable for their actions uphold the integrity of the research process. This helps prevent misconduct, such as data manipulation or unethical practices, which can compromise the validity and reliability of research findings.
- **Compliance with ethical standards**
  - Accountability ensures that researchers comply with ethical standards and legal requirements, preventing potential harm to participants and promoting ethical research practices.
- **Promoting accountability in the research community**
  - When researchers are accountable for their actions, it encourages a culture of ethical behaviour and responsibility within the research community. This fosters a positive environment for future research and collaboration.

Example of accountability in research ethics

In a clinical trial on a new medication:

- **Ethical conduct:** The researchers follow all ethical guidelines, including obtaining informed consent from participants and protecting their privacy throughout the study.
- **Transparency:** They disclose any potential conflicts of interest, such as funding from the pharmaceutical company producing the drug, and provide clear details on the trial design, methodology, and results.
- **Honesty in reporting:** When reporting results, the researchers accurately present both positive and negative outcomes, ensuring that no data is omitted or manipulated to create a biased conclusion.
- **Compliance:** The researchers comply with all IRB (IRB) regulations and national ethical standards for clinical trials.
- **Addressing misconduct:** If any issues of misconduct arise, such as participant safety concerns or data manipulation, the researchers are responsible for investigating the issue and taking corrective action, such as halting the trial or notifying regulatory bodies.

Best practices for ensuring accountability in research

- **Clear documentation**
  - Keep detailed records of all aspects of the research process, including informed consent forms, data collection procedures, and any changes to the study protocol. This ensures that the research can be reviewed and verified.
- **Adherence to ethical guidelines**
  - Follow established ethical guidelines, such as those set by IRBs or professional associations, to ensure the research is conducted responsibly.

- **Transparent reporting**
  - Be open about research findings, funding sources, and any potential conflicts of interest. Publish results in peer-reviewed journals and share data when possible to allow for verification and reproducibility.
- **Responding to concerns**
  - Be proactive in addressing any concerns or criticisms about the research. This includes responding to feedback from peers, participants, or oversight bodies and taking corrective action if necessary.
- **Ongoing education**
  - Researchers should stay informed about ethical standards and best practices in their field through ongoing education and training. This helps maintain accountability and ethical integrity throughout their careers.
- **Collaboration and peer review**
  - Engage in collaborative research and submit work for peer review to ensure that the research is held to high standards of accountability and scrutiny.

2.6.1.7 Justice

Justice in research ethics refers to the principle of fairness in the distribution of the benefits and burdens of research. It ensures that all individuals and groups are treated equitably, with no one group being unfairly burdened or excluded from the benefits of research. Researchers must ensure that participants are selected fairly, that their rights are respected, and that the benefits of the research are shared justly.

Key aspects of justice in research ethics

- **Fair distribution of risks and benefits**
  - Justice ensures that the risks and benefits of research are equitably distributed. Researchers should avoid exploiting vulnerable populations and ensure that any burdens associated with the research are not unfairly placed on one group while others benefit.
- **Equitable selection of participants**
  - Participants should be selected for research studies based on the scientific needs of the study rather than on convenience, such as choosing participants who are easily accessible. All individuals should have equal opportunity to participate in research if they meet the study's criteria.
- **Avoiding exploitation**
  - Researchers must avoid exploiting vulnerable groups (e.g., children, economically disadvantaged individuals, or marginalized populations). Special protections should be in place for such groups to ensure they are not unduly coerced or subjected to unfair risks.

- **Respecting diversity and inclusivity**
    - Research should include diverse populations to ensure that the findings are generalizable and applicable to different groups. Justice requires that research not exclude certain populations without valid scientific reasons, such as race, gender, or socioeconomic status.
- **Transparency and accountability**
    - Justice requires that researchers be transparent in their intentions and actions. They should ensure that participants are informed about the potential benefits and risks of the study, and they should be held accountable for how the research is conducted.
- **Access to benefits**
    - The benefits of research, such as new treatments, technologies, or knowledge, should be accessible to all, especially the populations who may have contributed to the research. Researchers and institutions have an ethical obligation to ensure that the results of research are used to benefit society as a whole.

Importance of justice in research ethics

- **Promotes fairness**
    - The principle of justice ensures that research is conducted fairly, preventing exploitation and ensuring that no group is unfairly burdened or excluded from the benefits of research.
- **Protects vulnerable populations**
    - Justice protects vulnerable populations by ensuring that they are not disproportionately exposed to risks without corresponding benefits. This helps prevent exploitation and ensures that vulnerable groups are treated with respect.
- **Fosters trust**
    - When research is conducted justly, it fosters trust between researchers, participants, and society. People are more likely to engage in research and support its findings if they believe the process is fair and equitable.
- **Enhances research integrity**
    - Ensuring justice in research helps maintain the integrity of the research process. By ensuring fair treatment of participants and equitable distribution of research benefits, researchers uphold the ethical standards of the field.

Example of justice in research ethics

In a clinical trial for a new vaccine:

- **Equitable selection**: Participants are chosen from a diverse range of demographics, including various ages, races, genders, and socioeconomic backgrounds, to ensure the results are applicable to the general population.

- **Fair distribution of risks and benefits:** The trial is designed to minimize risks, and the benefits, such as access to the new vaccine, are made available to all participants. Participants who are part of the control group are offered the vaccine after the trial concludes, ensuring they are not left without benefit.
- **Informed consent:** All participants are fully informed about the study's potential risks and benefits, and their consent is obtained voluntarily, without coercion.

Best practices for ensuring justice in research

- **Fair recruitment practices**
  - Researchers should ensure that participant recruitment is based on scientific criteria and not on convenience or exploitation. This involves actively seeking participants from diverse and underrepresented groups to ensure fairness.
- **Inclusive research design**
  - Research designs should be inclusive, considering how findings may apply to different populations. Exclusion of certain groups should only occur if there is a valid scientific reason, such as safety concerns.
- **Minimizing risk for vulnerable populations**
  - Extra care should be taken when involving vulnerable groups in research. Ethical guidelines should be followed to ensure their protection, and they should be fully informed of the risks and benefits of participating.
- **Equitable access to research findings**
  - Researchers and institutions should ensure that the benefits of research, such as medical treatments or scientific advancements, are made accessible to the populations that may benefit most from them.
- **Transparent reporting**
  - Researchers should report findings in a transparent and accessible manner, ensuring that the results of research are shared equitably with the public and that no group is excluded from access to the knowledge gained.

2.6.2 Importance of research ethics

2.6.2.1 Protects participants

Research ethics plays a critical role in protecting the well-being, rights, and dignity of participants in research studies [49]. Ethical guidelines and principles ensure that research is conducted responsibly, minimizing harm and safeguarding participants from exploitation or unnecessary risks. Here's how research ethics protects participants:

Key ways research ethics protects participants

- **Informed consent**
  - One of the fundamental ethical principles is ensuring that participants voluntarily agree to take part in research. Informed consent requires researchers to fully explain the study's purpose, procedures, risks, and benefits, allowing participants to make an informed decision about their involvement.
  - Participants must be made aware of their right to withdraw from the study at any time without facing negative consequences.
- **Confidentiality and privacy**
  - Research ethics mandates that participants' personal information and data be kept confidential. Ethical research practices ensure that sensitive information is protected and that participants' privacy is respected throughout the study.
  - Researchers are responsible for securely handling data and only sharing it in ways that protect participant identities.
- **Minimizing harm**
  - Research ethics require researchers to minimize potential physical, psychological, or emotional harm to participants. This involves carefully designing studies to avoid unnecessary risks and implementing measures to address any adverse effects that may arise.
  - Ethical guidelines emphasize that risks must be justified by the potential benefits of the research.
- **Protection of vulnerable populations**
  - Research ethics provides special protections for vulnerable populations, such as children, elderly individuals, or those with mental disabilities. These groups may be more susceptible to exploitation or harm, so additional safeguards are put in place to ensure their well-being.
  - Ethical research ensures that these populations are not unduly coerced into participation and that they receive appropriate care and support.
- **Right to withdraw**
  - Ethical research practices guarantee that participants can withdraw from a study at any point without fear of negative consequences. This principle protects participants from feeling trapped in a study and allows them to exercise autonomy over their involvement.
- **Debriefing**
  - After the completion of the study, especially in cases where deception is involved, participants must be debriefed. This means they are provided with full information about the study's true purpose and any relevant findings.

- Debriefing helps ensure that participants understand the study and its outcomes, and it provides them with an opportunity to ask questions and voice concerns.

Importance of protecting participants

- **Upholds ethical standards**
  - Protecting participants ensures that research is conducted according to high ethical standards, maintaining the integrity of the research process.
  - Ethical research promotes fairness, honesty, and transparency, ensuring that participants are treated with respect.
- **Builds trust**
  - When participants feel that their rights are respected and their well-being is protected, they are more likely to trust the research process. This trust is crucial for the success of the study and for future research involving human participants.
- **Promotes voluntary participation**
  - Ensuring that participants' autonomy is respected through informed consent and the right to withdraw promotes voluntary participation. This encourages greater engagement in research studies and ensures that participants are not coerced into taking part.
- **Ensures scientific integrity**
  - Ethical protections for participants lead to more reliable and accurate data. When participants are treated with respect and care, they are more likely to provide truthful and accurate responses, improving the validity of the study.
- **Legal and institutional compliance**
  - Research ethics help ensure compliance with legal and institutional regulations, protecting both participants and researchers from potential harm or legal repercussions.
  - Adhering to ethical standards also ensures that research findings are legally admissible and respected by the scientific community.

Example of protecting participants in research

In a medical trial for a new drug:

- **Informed consent:** Participants are fully informed of the trial's purpose, potential risks (e.g., side effects), and the benefits (e.g., possible access to new treatment). They voluntarily agree to participate.
- **Confidentiality:** Personal health information is securely stored and anonymized to protect participant privacy.
- **Minimizing harm:** The study is designed to minimize risks, and medical professionals are on hand to address any adverse effects that arise during the trial.

- **Vulnerable populations:** Extra precautions are taken if the study involves vulnerable groups, such as ensuring that children or elderly participants are properly supported and that their consent is obtained from a legal guardian if necessary.
- **Right to withdraw:** Participants can withdraw from the trial at any time without penalty, and they are reassured that they will continue to receive the best available care.
- **Debriefing:** After the trial, participants are informed of the study's outcomes, including whether the drug was successful or if there were any unforeseen risks.

2.6.2.2 Promotes trust

Trust is a foundational element of ethical research. When research is conducted with integrity, transparency, and respect for participants, it fosters trust between researchers, participants, and the broader community. Trust in research ethics ensures that participants are willing to engage in studies, share honest feedback, and contribute to scientific knowledge. It also builds confidence in the research findings and in the researchers themselves.

How research ethics promotes trust

- **Transparency**
    - Ethical research involves clear communication about the purpose, methods, risks, and benefits of a study. When researchers are transparent about their work, participants are more likely to trust the process and feel confident in their involvement.
    - Transparency in sharing research results, including both positive and negative outcomes, also contributes to trust by demonstrating honesty and accountability.
- **Respect for participants**
    - Research ethics emphasizes treating participants with dignity and respect, which helps build a trusting relationship. By ensuring that participants' rights, privacy, and autonomy are protected, researchers foster a sense of safety and respect.
    - Ensuring informed consent and the right to withdraw from a study without consequence reassures participants that their participation is voluntary and that their well-being is a priority.
- **Accountability**
    - Researchers who adhere to ethical standards and take responsibility for their actions promote trust by demonstrating that they are accountable for their conduct. This includes addressing any issues of misconduct, reporting results honestly, and correcting errors when necessary.

  - Accountability also extends to the handling of data, ensuring that it is collected, stored, and used responsibly, which builds trust in the integrity of the research process.
- **Protection of vulnerable populations**
  - Ethical guidelines provide special protections for vulnerable groups, ensuring they are not exploited or subjected to unnecessary risks. By prioritizing the safety and rights of these populations, researchers demonstrate their commitment to fairness and justice, which in turn fosters trust.
  - These protections also help avoid the history of exploitation in research, such as the unethical treatment of marginalized communities, reinforcing confidence in modern research practices.
- **Consistency and integrity**
  - Consistently applying ethical principles in all aspects of research, from design to reporting, establishes a reputation for integrity. Researchers who demonstrate ethical behaviour in every phase of the research process gain the trust of participants, peers, and the public.
  - Ethical research practices, such as avoiding plagiarism, fabrication, and falsification, ensure the reliability and validity of the research findings, contributing to the trustworthiness of the research.
- **Social responsibility**
  - Research that prioritizes ethical considerations, such as beneficence and justice, contributes to the public good. When researchers are seen as acting in the best interest of society and participants, they build trust within the community and demonstrate social responsibility.
  - Ethical research outcomes that benefit society, such as medical breakthroughs or environmental solutions, further reinforce trust in the research process and its role in improving lives.

Importance of trust in research ethics

- **Encourages participation**
  - When participants trust researchers and the research process, they are more likely to participate in studies, share honest information, and engage with the research. Trust reduces reluctance or fear about being manipulated or harmed.
  - Trust also encourages diverse participation, which leads to more comprehensive and accurate research findings.
- **Enhances research credibility**
  - Trust in the research process boosts the credibility of the research itself. When participants and the public believe that research is conducted ethically and responsibly, the findings are more likely to be taken seriously and respected by the scientific community and beyond.

- **Supports long-term relationships**
  - Ethical research practices build long-term relationships between researchers, participants, and the community. When participants feel respected and protected, they are more likely to participate in future studies and recommend others to do so, contributing to the growth of research efforts.
- **Promotes positive reputation**
  - Institutions and researchers who consistently adhere to ethical guidelines earn a positive reputation. This reputation attracts more funding, partnerships, and opportunities for future research.
  - A strong ethical reputation also leads to greater public confidence in the research outcomes, particularly when the results are made publicly available.
- **Reduces research bias**
  - Trust also plays a role in reducing bias. When researchers adhere to ethical guidelines and operate transparently, it minimizes the risk of manipulation, selective reporting, or other forms of bias that can undermine the trustworthiness of the findings.

Example of promoting trust in research ethics

In a public health study on a new vaccination:

- **Transparency:** Researchers clearly explain the purpose of the study, the potential risks, and the expected benefits of the vaccine. They also inform participants of any possible side effects and disclose any funding sources or conflicts of interest.
- **Respect for participants:** Participants are treated with dignity, their privacy is protected, and they are fully informed about their right to withdraw from the study at any time.
- **Accountability:** Researchers are open about the results, whether positive or negative, and address any concerns raised during the study, such as adverse reactions. They ensure that the data is handled responsibly and reported honestly.
- **Protection of vulnerable groups:** Extra precautions are taken for vulnerable populations, such as elderly participants, ensuring they are adequately informed and protected from undue risks.
- **Consistency and integrity:** The researchers follow ethical protocols consistently, ensuring the integrity of the study and building trust in their findings.

2.6.2.3 Enhances credibility

Credibility in research refers to the trustworthiness and reliability of the research process and its findings. Ethical practices in research are crucial in enhancing this credibility, ensuring that the research is conducted with integrity and that the results are accurate, valid, and trustworthy. By adhering to ethical guidelines, researchers demonstrate their commitment to producing credible and high-quality work.

How research ethics enhances credibility

- **Ensures validity and accuracy**
  - Ethical research practices, such as honest reporting of data and results, help ensure that the findings are accurate and valid. Researchers who follow ethical guidelines avoid practices like falsification or manipulation of data, which can undermine the credibility of their work.
  - Ethical research design also involves proper sampling, data collection, and analysis methods, which contribute to the reliability and validity of the findings.
- **Prevents bias**
  - Ethical standards require researchers to minimize bias in their work. This includes avoiding conflicts of interest, ensuring transparency in data collection and analysis, and reporting all results, even if they do not align with the researchers' expectations or hypotheses.
  - By eliminating bias, ethical research ensures that the results are a true reflection of the research question, thus enhancing the credibility of the findings.
- **Promotes transparency**
  - Ethical research practices require researchers to be transparent about their methods, data, and results. This transparency allows other researchers and the public to critically evaluate the research, replicate studies, and verify the findings.
  - Transparency also includes disclosing any potential conflicts of interest or funding sources, which helps to ensure that the research is not unduly influenced by external pressures.
- **Ensures accurate representation of findings**
  - Ethical research requires that researchers accurately report their findings without selective reporting or omission of negative or inconvenient results. By presenting all data—both positive and negative—researchers demonstrate their commitment to honesty and integrity, which enhances the credibility of their work.
  - Researchers must avoid practices such as cherry-picking data or exaggerating results to fit preconceived notions or desired outcomes.
- **Fosters accountability**
  - Researchers who follow ethical guidelines are accountable for their actions and the results of their research. This accountability helps ensure that research is conducted responsibly and that researchers take responsibility for any mistakes or issues that arise.

    - Accountability also involves proper citation of sources and giving credit where it is due, ensuring that the work of others is respected and acknowledged, which further strengthens the credibility of the research.
- **Maintains public trust**
    - Ethical research practices build public trust in scientific inquiry and its outcomes. When researchers are seen as adhering to ethical principles, their findings are more likely to be accepted by the public, policymakers, and other stakeholders.
    - This trust is essential for the successful implementation of research findings, particularly in fields such as healthcare, environmental science, and social policy.
- **Prevents misconduct**
    - Ethical research practices help prevent research misconduct, such as plagiarism, data fabrication, and falsification. By maintaining high ethical standards, researchers ensure that their work is credible and that their reputation, as well as the integrity of the field, is upheld.

Importance of enhancing credibility through research ethics

- **Increases acceptance of research findings**
    - When research is conducted ethically, it is more likely to be accepted and respected by the scientific community and the public. Ethical practices ensure that the findings are based on solid evidence, which increases their influence and impact.
- **Strengthens the research process**
    - Ethical research practices improve the overall quality of the research process by ensuring that the methods are sound, the data is reliable, and the results are valid. This strengthens the overall integrity of the research field and contributes to the advancement of knowledge.
- **Promotes replicability**
    - Credible research is replicable, meaning that other researchers can follow the same methods and obtain similar results. Ethical research practices ensure that the methodology is clearly documented and accessible, which supports the replicability of the study and enhances its credibility.
- **Supports evidence-based decision-making**
    - Credible research is essential for evidence-based decision-making, particularly in areas such as healthcare, policy, and education. When research is conducted ethically, it provides a solid foundation for making informed decisions that can benefit society.

Example of enhancing credibility in research ethics

In a clinical trial testing a new medication:

- **Valid and accurate data:** The researchers collect and analyze data using scientifically sound methods. They report both the positive effects and any side effects or negative outcomes, ensuring an honest representation of the results.
- **Prevention of bias:** The researchers ensure that the study design is free from bias by randomizing participants and controlling for confounding variables. They disclose any potential conflicts of interest, such as funding from the pharmaceutical company.
- **Transparency:** The researchers provide detailed information about their methods, data collection procedures, and statistical analyses, allowing others to evaluate the study's reliability and validity.
- **Accountability:** The researchers are transparent about any issues that arise during the trial, such as unexpected side effects, and they take responsibility for addressing them.

2.7 EFFECTIVE TECHNICAL WRITING

Technical writing conveys complex information in a clear, concise, and structured manner. Begin with a strong outline to organize your ideas. Use precise language and avoid jargon where possible [50]. Ensure logical flow between sections using headings and subheadings. Include visuals such as charts and diagrams to support your arguments. Revise and proofread thoroughly to eliminate errors and enhance clarity.

2.7.1 Technical writing skills encompass

2.7.1.1 Understanding the audience

Understanding the audience means identifying who will read your document and tailoring the content, language, and presentation to meet their needs, expectations, and level of expertise [51].

Importance

- Ensures the content is relevant and accessible.
- Avoids overloading readers with unnecessary details or jargon.
- Builds trust and credibility by addressing the audience's specific concerns or questions.

Key steps

- **Identify the audience:** Determine if the readers are experts, non-experts, or a mixed group.

- **Assess their knowledge level:** Understand what they already know and what they need to learn.
- **Define their purpose:** Consider why they are reading the document—seeking information, instructions, or decision-making guidance.
- **Adapt language and tone:** Use terminology appropriate for their familiarity with the subject.
- **Focus on relevance:** Highlight information that directly addresses their needs or goals.

Example

- Writing a technical manual for engineers may include detailed specifications and technical jargon.
- Preparing a summary for stakeholders or clients would focus on high-level insights and implications, avoiding overly technical language.

2.7.1.2 Writing for clarity

Writing for clarity involves presenting information in a straightforward, unambiguous, and easily understandable manner, ensuring readers can grasp the intended message without confusion [52].

Importance

- Prevents misunderstandings and errors.
- Enhances the efficiency of communication.
- Builds trust and credibility with the audience.

Key practices

- **Use simple language:** Avoid unnecessary jargon or complex vocabulary unless required for the audience.
- **Be concise:** Eliminate redundant words or phrases while retaining the core message.
- **Define technical terms:** Provide clear definitions for any specialized terms.
- **Structure sentences logically:** Use active voice and maintain a subject-verb-object order where possible.
- **Break down complex ideas:** Use lists, bullet points, or subsections to simplify intricate concepts.
- **Avoid ambiguity:** Use precise terms and avoid phrases that could be interpreted in multiple ways.
- **Provide examples:** Illustrate concepts with relevant examples to enhance understanding.

Example

Unclear: "Ensure all necessary components are accounted for before initiating operations."
Clear: "Before starting operations, verify that all required components, such as screws, bolts, and cables, are present."

Benefits

- Readers quickly understand the content without additional explanations.
- Reduces the risk of errors in implementation or interpretation.
- Saves time for both the writer and the reader.

2.7.1.3 Using good grammar and punctuation

Good grammar and punctuation involve adhering to the established rules of language to ensure written communication is clear, professional, and error-free.

Importance

- Enhances readability and comprehension.
- Demonstrates professionalism and attention to detail.
- Avoids potential misinterpretations caused by grammatical errors.

Key practices

- **Follow basic grammar rules**
  1. Ensure subject–verb agreement.
  2. Use correct tense and maintain consistency.
  3. Avoid misplaced or dangling modifiers.
- **Punctuate correctly**
  1. Use commas to separate clauses and avoid run-on sentences.
  2. Place periods, question marks, and exclamation points appropriately.
  3. Use colons and semicolons to organize lists or connect related ideas.
- **Avoid common mistakes**
  1. Misusing homophones (e.g., their/there/they're).
  2. Mixing up possessives and contractions (e.g., it's vs. its).
  3. Improper capitalization or punctuation in titles.
- **Proofread for errors**
  1. Re-read the text to catch typos, misplaced punctuation, or grammatical mistakes.

2. Use tools like grammar checkers to supplement manual proofreading.
- **Be consistent**
1. Maintain a uniform tone and style throughout the document.
2. Use consistent punctuation for lists, headings, and formatting.

Example

Incorrect: "The data, which was collected over several months are highly reliable."
Correct: "The data, which were collected over several months, are highly reliable."

Benefits

- Increases the professionalism of the document.
- Ensures the message is interpreted as intended.
- Builds trust with the audience by showcasing meticulousness.

2.7.1.4 Being consistent

Consistency in technical writing refers to maintaining uniformity in style, terminology, tone, formatting, and structure throughout a document or series of documents.

Importance

- Enhances readability and comprehension.
- Builds trust by providing a professional and cohesive presentation.
- Reduces confusion, especially in complex or technical content.

Key areas of consistency

- **Terminology**
  1. Use the same terms for the same concepts throughout the document.
  2. Avoid switching between synonyms or abbreviations unless clearly defined.
  3. Example: If "Unmanned Aerial Systems" is used, avoid alternating with "UAS" without clarification.
- **Formatting**
  1. Standardize font types, sizes, and spacing.
  2. Ensure uniformity in headings, subheadings, bullet points, and numbering styles.
  3. Apply consistent styles for tables, figures, and captions.
- **Tone and Style**
  1. Maintain a formal or professional tone unless otherwise specified.
  2. Avoid shifts in voice (e.g., from passive to active) unless intentional for clarity.

- **Referencing**
  1. Follow the same citation style (e.g., APA and MLA) throughout.
  2. Consistently format in-text citations, footnotes, and bibliographies.
- **Units and measurements**
  1. Use a consistent system (e.g., metric or imperial) and format (e.g., cm or centimetres).
- **Grammar and punctuation**
  1. Ensure consistent use of Oxford commas, hyphenation, and capitalization rules.
- **Document structure**
  1. Follow the same layout for similar sections (e.g., introduction, methodology, and conclusions).

Example

Inconsistent
"The software update will improve efficiency."
"This update improves system performance and efficiency."

Consistent
- "The software update will improve system performance and efficiency."

Tips for maintaining consistency

- Create and adhere to a style guide or template.
- Use tools like document management software to track formatting.
- Review the document holistically to identify and correct inconsistencies.

Benefits

- Improves the user's ability to follow and understand the content.
- Ensures a professional and polished presentation.
- Saves time in editing and reviewing processes.

2.7.1.5 Having a logical flow

Logical flow refers to the clear and systematic organization of information in a document, ensuring that ideas are presented in a sequence that is easy for the reader to follow and understand.

Importance

- Enhances readability by guiding the reader through the content.
- Improves comprehension by connecting ideas seamlessly.
- Helps maintain the reader's interest and focus.

Key principles for achieving logical flow

- **Structured organization**
  1. Use a clear outline with headings and subheadings to break down information.
  2. Arrange sections in a natural progression, such as chronological order, problem-solution format, or general-to-specific.
- **Clear transitions**
  1. Use transition words and phrases (e.g., "therefore," "however," and "next") to link ideas and sections.
  2. Ensure each paragraph leads naturally to the next.
- **Focused paragraphs**
  1. Begin each paragraph with a topic sentence that summarizes its main idea.
  2. Keep paragraphs concise and focused on a single concept or point.
- **Logical relationships**
  1. Group related ideas together to avoid jumping between topics.
  2. Use bullet points or numbered lists for clarity when presenting multiple items.
- **Introduction and conclusion**
  1. Start with an introduction that outlines the purpose and scope of the document.
  2. End with a conclusion that summarizes key points and reinforces the main message.

Example

Poor Flow

- Describes results without introducing the methodology.
- Discusses recommendations before presenting findings.

Improved Flow

- **Introduction:** State the purpose and background of the document.
- **Methodology:** Explain the approach used.
- **Findings:** Present the results logically.
- **Discussion:** Interpret the findings.
- **Conclusion and Recommendations:** Summarize and provide actionable insights.

Tips for maintaining logical flow

- Use outlines or mind maps during the planning phase.
- Read the document aloud to identify any abrupt transitions or gaps.
- Get feedback from others to ensure clarity and coherence.

Benefits

- Improves the reader's understanding and retention of information.
- Reduces confusion and misinterpretation.
- Ensures the document achieves its intended purpose effectively.

2.7.1.6 Making sure all information is correct

Ensuring all information in a technical document is accurate, up-to-date, and relevant to the topic being discussed.

Importance

- Builds credibility and trust with the audience.
- Prevents misunderstandings or misuse of information.
- Upholds professional and ethical standards in communication.

Key strategies for ensuring accuracy

- **Thorough research**
  1. Use reliable and authoritative sources, such as peer-reviewed journals, official reports, and credible websites.
  2. Cross-check information across multiple sources to confirm its validity.
- **Fact-checking**
  1. Verify numerical data, technical specifications, and terminology.
  2. Ensure all examples, case studies, or real-world references are accurate and applicable.
- **Stay updated**
  1. Use the most recent and relevant information, especially in fields where knowledge evolves rapidly.
  2. Regularly review and update documents to reflect current standards and practices.
- **Subject matter expert (SME) review**
  1. Consult with experts in the field to validate technical content.
  2. Incorporate feedback to refine and correct any errors.
- **Proofreading and testing**
  1. Check for errors in calculations, citations, and logical arguments.
  2. If applicable, test instructions, processes, or designs described in the document to ensure they work as intended.
- **Avoid assumptions**
  1. Base all claims and statements on evidence, not personal beliefs or assumptions.
  2. Clearly indicate if any information is based on estimates or pending confirmation.

Example of incorrect vs. correct information:

Incorrect: Stating that "the system supports 1,000 concurrent users" without testing or documentation.

Correct: "Based on load-testing results, the system supports up to 1,000 concurrent users under optimal conditions."

Consequences of incorrect information

- Misleading readers and causing errors in implementation.
- Damaging professional reputation and credibility.
- Potential legal or ethical repercussions.

Tips for ensuring correctness

- Create a checklist to verify critical data before finalizing the document.
- Use tools like citation managers and data validation software.
- Encourage peer reviews and audits of the document.

Benefits

- Enhances the reliability and utility of the document.
- Strengthens the audience's confidence in the writer's expertise.
- Reduces the risk of costly mistakes or miscommunication.

2.7.1.7 *Including enough detail so readers won't have to look up anything*

Providing comprehensive and precise information within a document so that readers can fully understand the content without needing to consult external sources.

Importance

- Enhances **reader efficiency** by reducing time spent searching for additional information.
- Builds **trust** in the document's reliability and completeness.
- Ensures the document meets its purpose as a standalone resource.

Key strategies for including enough detail

- **Understand the audience**
  - Determine the reader's level of expertise.
  - Include foundational information for novices and detailed specifics for experts.
- **Anticipate questions**
  - Think about what the audience might ask while reading.
  - Address common queries or ambiguities preemptively in the content.
- **Be explicit**
  - Avoid vague terms and generalizations.
  - Use precise language, examples, and clear definitions for technical terms.

- **Use supporting elements**
  - Include diagrams, charts, and tables to explain complex concepts.
  - Provide step-by-step instructions or processes when applicable.
- **Provide context**
  - Explain why certain methods, materials, or decisions were chosen.
  - Include background information to help readers grasp the broader significance.
- **Cite and explain**
  - If external sources are referenced, summarize their key points within the document.
  - Offer a brief explanation rather than just providing a link or citation.

Examples of sufficient vs. insufficient detail

Insufficient detail
"Follow the standard procedure for calibration."

Sufficient detail
"To calibrate the instrument, use a 10mL pipette to add the buffer solution, ensuring the pH is adjusted to 7.0 ± 0.1. Then, run the calibration software provided with version 3.2.1 or later."

Benefits of including enough detail

- Prevents errors caused by assumptions or missing steps.
- Increases the usability and value of the document.
- Improves the reader's confidence in implementing or understanding the content.

Potential challenges

- Balancing detail with conciseness to avoid overwhelming the reader.
- Ensuring that all included details are relevant and up-to-date.

Tips for success

- Use bullet points, numbered lists, or headings to organize detailed content clearly.
- Test the document by having someone unfamiliar with the topic read it and check for gaps in understanding.
- Regularly review and revise to ensure completeness and relevance.

Including enough detail ensures that a technical document serves as a reliable, comprehensive resource, empowering readers to use the information effectively without seeking external clarification.

2.7.1.8 Being organized

The practice of structuring information logically and systematically to enhance readability and comprehension in technical documents.

Importance

- Facilitates easy navigation through the document.
- Helps readers locate specific information quickly.
- Ensures a seamless flow of ideas, reducing confusion.
- Reflects professionalism and attention to detail.

Key strategies for being organized

- **Outline before writing**
  - Create a detailed framework of the document's sections and subsections.
  - Arrange topics in a logical sequence, such as chronological, thematic, or hierarchical order.
- **Use headings and subheadings**
  - Divide content into manageable sections with clear, descriptive titles.
  - Maintain a consistent hierarchy for headings (e.g., H1 for main topics and H2 for subtopics).
- **Maintain logical flow**
  - Ensure that one idea leads naturally to the next.
  - Use transitional phrases to connect sections and concepts.
- **Prioritize information**
  - Present the most critical information first (e.g., executive summaries or key findings).
  - Place supplementary details, such as references or appendices, at the end.
- **Use lists and tables**
  - Break down complex ideas into bullet points or numbered lists for clarity.
  - Present data in tables or charts to make comparisons and patterns more accessible.
- **Provide navigation aids**
  - Include a table of contents, page numbers, and an index for lengthy documents.
  - Use hyperlinks or bookmarks in digital formats for quick access to specific sections.
- **Review and revise**
  - Reread the document to ensure the structure supports clarity and logical progression.
  - Seek feedback from others to identify potential areas of confusion.

Examples of organization in technical writing

Poorly organized

- Information is scattered, with no clear headings or sequence.
- Data is embedded in lengthy paragraphs, making it hard to locate.

Well-organized

- The document starts with an introduction and purpose.
- Key topics are divided into sections with clear headings.
- Data is presented in tables, and a summary concludes the document.

Benefits of being organized

- Enhances reader engagement and retention.
- Minimizes the risk of misinterpretation or overlooking critical details.
- Saves time for both the writer and the reader.

Potential challenges

- Over-organizing can make the document appear rigid or fragmented.
- Balancing structure with flexibility to adapt to diverse content.

Tips for success

- Keep the audience's needs in mind when organizing content.
- Use software tools like word processors or document management systems to assist with formatting and layout.
- Regularly update the organization as the document evolves.

Being organized is essential in technical writing to ensure that complex information is presented in a clear, coherent, and accessible manner, enabling readers to derive maximum value with minimal effort.

2.7.1.9 Follow the organization's standards and guidelines

Adhering to the established guidelines, style manuals, and protocols set by your organization when creating technical documents.

Importance

- Ensures uniformity and professionalism across all documents.
- Reflects the organization's brand identity and values.
- Facilitates easy comprehension and usability for internal and external stakeholders.

Key aspects of organizational standards

- **Style guides**
    - Follow the prescribed format (e.g., APA, MLA, IEEE, or in-house style manuals).

    - Use specific templates for reports, proposals, or other document types.
- **Tone and language**
    - Maintain a consistent tone (e.g., formal, neutral, or technical).
    - Use terminology aligned with organizational or industry-specific jargon.
- **Branding elements**
    - Incorporate the organization's logo, colours, and fonts as required.
    - Use pre-approved layouts and designs for headers, footers, and title pages.
- **Document structure**
    - Follow the standard structure for specific document types, such as including an executive summary, methodology, or conclusion sections.
    - Adhere to page numbering, margin sizes, and alignment rules.
- **Formatting rules**
    - Use consistent headings, bullet styles, and indentation.
    - Ensure proper spacing, font sizes, and alignment as per the organization's preferences.
- **Legal and ethical compliance**
    - Include necessary disclaimers, confidentiality notices, or copyright statements.
    - Follow data privacy regulations and intellectual property policies.
- **Quality control**
    - Ensure documents are reviewed and approved through the organization's specified processes.
    - Conduct thorough proofreading to meet the organization's standards for accuracy and quality.

Examples of organizational standards in practice

Scenario 1: A multinational company may require all internal reports to use the Arial font, size 11, with 1-inch margins, and include a cover page with the company logo.

Scenario 2: A research institution may mandate that all technical papers follow IEEE formatting guidelines, with specific rules for citations, figures, and references.

Benefits of following organizational standards

- Promotes consistency and professionalism in communication.
- Simplifies document review and approval processes.
- Builds trust and credibility with stakeholders.

Potential challenges

- Adjusting to new standards when transitioning to a different organization.

- Ensuring compliance across teams with diverse writing styles.
- Keeping up-to-date with evolving organizational guidelines.

Tips for success

- Familiarize yourself with your organization's style guides and templates early on.
- Seek clarification from supervisors or colleagues if standards are unclear.
- Use document creation tools that integrate organizational standards, such as pre-formatted templates.
- Regularly check for updates to guidelines or protocols.

By following organizational standards, technical writers contribute to a cohesive and professional representation of their organization's communication, ensuring that all documents meet the desired quality and branding expectations.

2.7.1.10 Knowing when to use tables and figures

The ability to discern when visual aids such as tables, charts, graphs, and images enhance understanding and when they might be unnecessary or counterproductive.

Purpose

Tables and figures simplify complex data, illustrate relationships, and provide visual emphasis. They are valuable tools for improving the clarity and impact of technical documents.

When to use tables

Use cases

- **Presenting numerical data:** To compare statistics, trends, or quantities.
- **Summarizing key information:** For concise presentation of data that would otherwise require lengthy explanations.
- **Providing structure:** To organize information into rows and columns for quick reference.

Example

A table listing the specifications of various drone models, including weight, battery life, and range, allows readers to easily compare features.

When to use figures (charts, graphs, and images)

Use cases

- **Visualizing trends:** Line graphs to show changes over time or bar charts for category comparisons.
- **Explaining processes:** Flowcharts to depict workflows or decision **trees.**
- **Highlighting relationships:** Scatterplots to show correlations between variables.
- **Enhancing understanding:** Diagrams or images to illustrate complex mechanisms or concepts.

Example

A pie chart representing the market share of drone manufacturers provides a quick snapshot of industry distribution.

When not to use tables and figures

Avoid

- **Redundancy:** If the same information is already explained clearly in the text.
- **Excessive complexity:** When visual aids are too detailed or confusing.
- **Irrelevance:** If the table or figure does not add value or is unrelated to the document's focus.

Example

A table with too many rows and columns, requiring extensive explanation, may confuse readers instead of clarifying the information.

Tips for effective use of tables and figures

- **Ensure relevance:** Only include visual aids that support or enhance the text.
- **Keep It simple:** Avoid overloading visuals with excessive data or annotations.
- **Label clearly:** Use descriptive titles, captions, and axis labels.
- **Provide context:** Explain the table or figure in the text and highlight key takeaways.
- **Maintain consistency:** Use a uniform style, font, and colour scheme throughout the document.
- **Choose the right type:** Select the most appropriate visual representation for the data or concept.

Benefits of using tables and figures effectively

- Enhances reader engagement and comprehension.

- Breaks up dense text, making documents more visually appealing.
- Communicates complex information succinctly.
- Strengthens the impact of technical documents.

By knowing when and how to use tables and figures, technical writers can effectively convey information, making their work both informative and accessible to diverse audiences.

2.7.2 Purpose of technical writing

The primary purpose of technical writing is to communicate complex information effectively and efficiently to a specific audience [53]. It aims to bridge the gap between technical knowledge and the user's understanding, ensuring clarity, accuracy, and usability [54]. The objectives can be summarized as follows:

2.7.2.1 Instruction

Instruction in technical writing refers to the process of clearly and systematically guiding the reader through specific tasks, procedures, or operations. It ensures that the audience can successfully achieve the intended outcome without confusion or error.

Purpose of instruction in technical writing

- **Facilitates understanding**
  Break down complex tasks into manageable steps, ensuring clarity and accessibility for diverse audiences.
- **Promotes efficiency**
  Provides concise and direct guidance, reducing time spent on trial and error.
- **Ensures accuracy**
  Prevents mistakes by offering detailed and precise directions, minimizing errors in implementation.
- **Enhances safety**
  Particularly in technical or industrial contexts, instructions highlight precautions to avoid accidents or misuse.
- **Supports learning**
  Educates users on processes, enabling them to build skills or operate tools independently.

Characteristics of effective instruction in technical writing

- **Clarity**: Avoids ambiguity and uses simple, straightforward language.
- **Step-by-step structure**: Break down processes into logical and sequential actions.
- **Consistency**: Uses consistent terminology and formatting throughout the instructions.

- **Visual aids:** Incorporates diagrams, flowcharts, or images to support textual guidance.
- **Contextual relevance:** Tailors instructions to the specific needs and expertise of the audience.

Examples of instruction in technical writing

- **User manuals:** Step-by-step guides for assembling furniture or operating machinery.
- **Software tutorials:** Instructions for installing or troubleshooting applications.
- **Safety procedures:** Guidelines for handling hazardous materials or operating heavy equipment.
- **Educational materials:** Laboratory manuals or e-learning content.

2.7.2.2 Explanation

Instruction in technical writing serves as a critical component to bridge the gap between information and action. It ensures that readers can perform specific tasks effectively, whether assembling a product, using software, or following safety protocols.

Why instructions are important

- **Guiding actions**
  Instructions translate technical knowledge into actionable steps. They provide users with the "how-to" needed to accomplish a task, making complex processes approachable.
- **Preventing errors**
  Clear and structured instructions reduce the likelihood of mistakes, ensuring tasks are performed accurately and efficiently.
- **Improving accessibility**
  Well-written instructions cater to a wide range of users, from novices to experts, enabling all to achieve the desired outcome without additional support.
- **Enhancing safety**
  Instructions often include warnings or precautions, safeguarding users from potential risks or hazards.

How instructions function in technical writing

- **Clarity and simplicity**
  They simplify technical jargon, ensuring that even non-experts can follow the process.
- **Sequential steps**
  Tasks are broken down into logical steps, enabling users to progress methodically.

- **Use of visuals**
  Images, diagrams, or flowcharts complement the text, making complex tasks easier to understand.
- **Focus on outcomes**
  Instructions are goal-oriented, guiding users towards achieving a specific result.

Real-world examples

- **Assembly manuals:** For furniture or electronics, instructions guide users through setup.
- **Technical guides:** Software installation or troubleshooting procedures.
- **Safety instructions:** Handling equipment or chemicals with proper precautions.
- **Educational resources:** Tutorials for students in labs or e-learning platforms.
- **Impact of clear instructions**
- When written effectively, instructions save time, reduce frustration, and foster user confidence. They transform technical complexity into actionable insights, ensuring that users can operate independently while adhering to safety and accuracy standards.

2.7.2.3 Documentation

Documentation is a cornerstone of technical writing, encompassing a range of materials designed to record, explain, and support the use or understanding of products, processes, or systems. It serves as a reference, providing detailed and accurate information to users, developers, or stakeholders.

Purpose of documentation

- **Knowledge preservation**
  Documentation ensures that critical information is recorded for future reference, reducing reliance on individual expertise.
- **User support**
  It helps users understand how to operate or interact with a product or system effectively.
- **Standardization**
  Documentation creates consistency by offering a single source of truth for processes, procedures, and systems.
- **Compliance**
  It fulfils regulatory or organizational requirements by maintaining accurate records of activities, designs, or procedures.

Types of documentation

- **User manuals**
  Guides for end-users, explaining product functionality, troubleshooting, and maintenance.
- **Technical manuals**
  Detailed information for developers, engineers, or technicians, including schematics, system requirements, and code documentation.
- **Process documentation**
  Step-by-step instructions for workflows, ensuring consistency and efficiency in organizational processes.
- **Policy documents**
  Formalized rules, regulations, or standards to guide actions and decision-making.
- **Training materials**
  Content designed to educate users, employees, or stakeholders about specific tools, systems, or concepts.

Characteristics of effective documentation

- **Clarity:** Written in simple, precise language for the intended audience.
- **Accuracy:** Free of errors and reflective of the latest information or processes.
- **Organization:** Logically structured, with sections, headings, and an index for easy navigation.
- **Completeness:** Covers all necessary aspects without leaving gaps in understanding.
- **Consistency:** Maintains uniform terminology, style, and formatting throughout.
- **Accessibility:** Available in formats suitable for the audience, such as digital, print, or interactive platforms.

Role of documentation in technical writing

- Documentation is not merely a record; it is a communication tool that bridges gaps in knowledge and supports decision-making. By adhering to best practices in documentation, technical writers ensure that users, developers, and stakeholders have the resources they need to succeed.

2.7.2.4 Persuasion

Persuasion in technical writing involves the strategic presentation of information to convince the audience to adopt a viewpoint, make a decision, or take action. Unlike purely informational writing, persuasive technical writing blends factual accuracy with compelling arguments tailored to the needs and concerns of the audience.

Purpose of persuasion in technical writing

- **Influencing decisions**
  To guide stakeholders, clients, or users towards adopting a solution, approving a proposal, or investing in a project.
- **Justifying actions**
  To provide evidence-based reasoning for implementing changes, adopting new systems, or following recommendations.
- **Encouraging compliance**
  To convince the audience to adhere to policies, procedures, or safety guidelines.
- **Securing resources**
  To advocate for funding, personnel, or other resources necessary for project success.

Key techniques for persuasive technical writing

- **Understand the audience**
  - Identify their needs, priorities, and potential objections.
  - Tailor the argument to resonate with their values and expectations.
- **Present clear benefits**
  - Highlight how the proposed idea, product, or solution addresses problems or adds value.
  - Use specific examples to demonstrate tangible outcomes.
- **Use data and evidence**
  - Support arguments with credible data, statistics, case studies, and expert opinions.
  - Ensure that evidence aligns with the audience's level of expertise.
- **Structure logically**
  - Begin with a strong introduction that outlines the purpose and stakes.
  - Progress through a logical sequence of arguments leading to a clear conclusion.
- **Appeal to emotion and logic**
  - Balance emotional appeals (e.g., safety and convenience) with logical reasoning (e.g., cost-effectiveness and efficiency).
- **Maintain professional tone**
  - Use clear and respectful language.
  - Avoid aggressive or overly subjective phrasing.

Examples of persuasive technical writing

- **Proposals**
  Documents advocating for project approval, funding, or strategic changes.

- **White papers**
  Detailed reports promoting the benefits of a product, service, or policy.
- **Policy briefs**
  Summaries aimed at influencing decision-makers to adopt or modify policies.
- **Instructional materials**
  Content that persuades users to follow specific safety or operational guidelines.

Characteristics of effective persuasive writing

- **Clarity:** Ensures the argument is easy to understand.
- **Credibility:** Builds trust through accuracy and authoritative sources.
- **Relevance:** Focuses on issues important to the audience.
- **Engagement:** Captures interest through compelling and relatable content.
- **Action-Oriented:** Encourages specific actions or decisions.

Role of persuasion in technical writing

- Persuasion transforms technical writing from mere information dissemination to an influential tool that drives decisions and actions. By integrating persuasive techniques, technical writers can effectively bridge the gap between data and decision-making, ensuring their work has a lasting impact.

2.7.2.5 Simplification

Simplification in technical writing focuses on making complex information easy to understand for a diverse audience. The goal is to break down intricate concepts, processes, or data into clear and straightforward language without losing essential details or accuracy.

Purpose of simplification

- **Improves accessibility:** Ensures the content is understandable for readers with varying levels of expertise.
- **Enhances comprehension:** Helps readers grasp key points quickly and accurately.
- **Saves time:** Reduces the effort needed to process and act on the information.
- **Encourages engagement:** Simplified content keeps readers interested and reduces frustration.

Techniques for simplification

- **Use plain language**
  - Avoid jargon, acronyms, or overly technical terms.

  - Replace complex words with simpler alternatives.
- **Organize content clearly**
  - Use headings, subheadings, and bullet points for structure.
  - Present ideas in a logical sequence.
- **Incorporate visual aids**
  - Use diagrams, tables, or charts to explain concepts visually.
  - Include labelled figures to reinforce written descriptions.
- **Break down processes**
  - Divide tasks or procedures into step-by-step instructions.
  - Provide examples or analogies to illustrate concepts.
- **Focus on key points**
  - Highlight essential information while omitting unnecessary details.
  - Use summaries or key takeaways for emphasis.
- **Test for readability**
  - Review content to ensure clarity and simplicity.
  - Use tools like readability tests to match audience expectations.

Examples of simplification

- **User manuals:** Instructions that explain how to use a product in easy-to-follow steps.
- **Quick start guides:** Condensed information for immediate application.
- **FAQs:** Direct answers to common questions.
- **Infographics:** Visual summaries of data or processes.

Benefits of simplification

- **Increased understanding:** Readers can act on the information confidently.
- **Reduced errors:** Clear instructions lower the risk of misinterpretation.
- **Broader audience reach:** Simplified content is accessible to both experts and non-experts.

Simplification is a cornerstone of effective technical writing, ensuring that complex ideas are conveyed in a way that serves the needs of all readers. By making content approachable, technical writers enhance its utility and impact.

2.8 STEPS FOR WRITING A REPORT

A well-structured report is essential for effectively communicating research findings. It ensures clarity, logical flow, and comprehensive documentation of your work. Below is a breakdown of the standard components of a report [54, 55]:

2.8.1 Title page

The title page is the first impression of your report and provides essential details to the reader. It should be professionally formatted and concise, including the following key elements:

Title of the report

- Clearly state the subject of the report.
- Use a descriptive and precise title that reflects the content of the report.
- Example: *"A Study on the Impact of Renewable Energy Adoption in Urban Areas."*

Author(s)

- List the name(s) of the individual(s) who authored the report.
- Include academic or professional titles, if applicable.

Affiliation

- Mention the institution, organization, or company the author(s) are associated with.
- Example: *Department of Mechanical Engineering, NIT Rourkela*

Date of submission

- Provide the date when the report is submitted.
- Format the date appropriately (e.g., December 19, 2024).

Optional Elements

- **Report number or code:** If applicable, include a unique identifier for the report.
- **Logo:** Add the logo of the affiliated institution or organization for a professional touch.
- **Supervisor's name:** If the report is part of an academic or professional project, include the name of the supervisor or guide.

Purpose

- The title page serves as a quick reference for the reader to identify the report and its origin. It ensures that all formal details are prominently displayed for academic or professional documentation.

Formatting tips

- Centre the content for a balanced and clean layout.
- Use consistent font style and size, typically Times New Roman or Arial, with a font size of 12–14 points for the main text.

- Ensure proper alignment and spacing to maintain a professional appearance.

2.8.2 Abstract

The abstract is a concise summary of the entire report, providing an overview of its objectives, methodology, key findings, and conclusions. It allows readers to quickly understand the report's purpose and significance.

Key components of an abstract

- **Objectives**
  - Clearly state the purpose of the study or the problem being addressed.
  - Example: *This report investigates the effectiveness of renewable energy systems in reducing urban carbon footprints.*
- **Methodology**
  - Briefly describe the approach, techniques, or tools used in the research.
  - Example: *A combination of field surveys, simulation models, and data analysis was employed to assess energy consumption patterns.*
- Key Findings
  - Summarize the main results or outcomes of the research.
  - Example: *The findings indicate a 30% reduction in energy costs and a 25% decrease in emissions with the adoption of solar and wind energy systems.*
- Conclusions
  Highlight the implications or significance of the results
  - Example: *The study demonstrates the viability of integrating renewable energy in urban planning to achieve sustainable development goals.*

Purpose of an abstract

- Acts as a standalone summary for readers to decide whether the report is relevant to their interests.
- Helps in indexing and retrieval for academic and professional databases.

Formatting tips

- Limit the abstract to 150–250 words.
- Avoid technical jargon or overly complex language.
- Write in a single, cohesive paragraph.
- Use past tense for completed work and present tense for general statements or conclusions.

The abstract serves as the gateway to your report, setting the tone and providing a snapshot of its content. It is often the first section read and should be compelling yet precise.

2.8.3 Introduction

The introduction is the opening section of a report, designed to provide context and establish the foundation for the research. It introduces the topic, defines the problem, and explains the significance of the study.

Key components of an introduction

- **Background information**
  - Provide an overview of the topic to familiarize the reader.
  - Example: *Renewable energy has emerged as a critical solution to combat climate change and reduce dependence on fossil fuels.*
- **Problem definition**
  - Clearly state the research problem or question being addressed.
  - Example: *Despite advancements, the integration of renewable energy in urban settings remains challenging due to infrastructure and policy limitations.*
- **Research objectives**
  - Outline the specific goals or aims of the study.
  - Example: *This study aims to evaluate the feasibility and impact of renewable energy systems in reducing urban carbon footprints.*
- **Significance of the study**
  - Explain why the research is important and its potential contributions.
  - Example: *The findings of this research will guide policymakers and urban planners in adopting sustainable energy practices.*
- **Scope of the study**
  - Briefly mention the boundaries or limitations of the research.
  - Example: *The study focuses on mid-sized urban areas in the context of developing countries.*

Purpose of an introduction

- Captures the reader's interest and provides a roadmap for the report.
- Sets the stage for understanding the relevance and direction of the research.

Formatting tips

- Begin with a compelling opening sentence to draw the reader's attention.
- Use clear and concise language.

- Keep the introduction focused and avoid unnecessary details.
- Transition smoothly into the methodology or background sections.

A well-crafted introduction ensures the reader understands the purpose and context of the report, making them more engaged with the subsequent sections.

2.8.4 Methodology

The methodology section outlines the approach, tools, and techniques used to conduct the research. It provides a clear and detailed description of how the study was carried out, ensuring transparency and reproducibility.

Key components of methodology

- **Research design**
  - Specify the type of research conducted (e.g., experimental, observational, and survey-based).
  - Example: *This study employs a mixed-methods approach, combining qualitative interviews with quantitative data analysis.*
- **Data collection methods**
  - Detail how data was gathered (e.g., surveys, experiments, and archival research).
  - Example: *Data was collected through structured interviews with urban planners and surveys distributed to local residents.*
- **Instruments and tools**
  - Mention any equipment, software, or tools used for data collection or analysis.
  - Example: *Geographic Information Systems (GIS) were utilized to map renewable energy adoption patterns.*
- **Sampling techniques**
  - Describe the population or sample and the method used to select participants.
  - Example: *A stratified random sampling technique was employed to ensure representation across different income groups.*
- **Procedures**
  - Provide a step-by-step explanation of the process followed.
  - Example:
    1. *Pilot surveys were conducted to refine the questionnaire.*
    2. *Data collection spanned three months, involving field visits and online surveys.*
- **Data analysis methods**
  - Explain the statistical or analytical techniques used to interpret the data.
  - Example: *Regression analysis was conducted to identify correlations between policy incentives and renewable energy adoption.*

- **Ethical considerations**
  - Highlight measures taken to ensure ethical research practices.
  - Example: *Informed consent was obtained from all participants, and their anonymity was maintained.*

Purpose of the methodology section

- Enables readers to understand and replicate the study.
- Demonstrates the rigour and reliability of the research process.

Formatting tips

- Use subheadings to organize content clearly.
- Provide sufficient detail without overwhelming the reader with unnecessary technical jargon.
- Use diagrams or flowcharts if needed to illustrate complex processes.

A robust methodology ensures that the research findings are credible, reproducible, and grounded in a systematic approach.

2.8.5 Results

The results section presents the key findings of your research in a clear and organized manner. It focuses on the data and observations without interpretation or commentary, leaving the analysis for the discussion section.

Key components of the results section

- **Objective presentation of data**
  - Provide the findings as they were observed, without bias or opinion.
  - Example: *The survey revealed that 65% of participants favoured renewable energy incentives, while 20% opposed them.*
- **Use of tables and figures**
  - Use visual aids like charts, graphs, and tables to summarize complex data effectively.
- **Descriptive statistics**
  - Include means, medians, standard deviations, and other statistical summaries if applicable.
  - Example: *The average annual energy savings for households adopting solar panels was $1,200 (SD = $300).*
- **Highlighting significant findings**
  - Emphasize key results that directly address your research objectives.
  - Example: *Regression analysis showed a significant positive correlation ($r = 0.85$) between government subsidies and solar panel adoption rates.*

- **Comparative results**
  - Present comparisons between different groups, conditions, or variables.
  - Example: *Urban households were 30% more likely to adopt renewable energy than rural households.*
- **Avoid interpretation**
  - Refrain from analyzing or discussing the implications of the findings. Save this for the discussion section.
  - Example: *While adoption rates were higher in urban areas, this trend was consistent across all age groups.*

Purpose of the results section

- To provide a factual account of what was discovered during the research.
- To serve as the foundation for interpretation and discussion in subsequent sections.

Formatting tips

- Use numbered headings or subheadings to organize results by research objectives or themes.
- Label all tables and figures with descriptive titles and include legends if needed.
- Ensure all visual aids are clear, properly formatted, and referenced in the text.

By maintaining clarity, conciseness, and objectivity, the results section ensures that readers can fully understand and trust the data generated by the research.

2.8.6 Discussion

The discussion section interprets and analyzes the results of your research. It connects the findings to your objectives, compares them with previous studies, and explores their implications. This section provides the opportunity to explain the significance of your work and suggest areas for future research.

Key components of the discussion section

- **Interpretation of results**
  - Explain the meaning of the findings and how they address the research objectives.
  - Example: *The high adoption rate of renewable energy in urban areas suggests that accessibility to resources plays a crucial role in influencing consumer behaviour.*

- **Comparison with existing literature**
  - Relate your findings to those of previous studies, highlighting similarities, differences, or contradictions.
  - Example: *Unlike Smith et al. (2021), who reported minimal rural adoption, our study shows moderate adoption driven by government incentives.*
- **Significance of findings**
  - Emphasize the broader implications of your results for the field or practical applications.
  - Example: *These findings underscore the importance of targeted subsidies to encourage sustainable energy practices across diverse demographics.*
- **Consideration of limitations**
  - Acknowledge any limitations in your research and their potential impact on the results.
  - Example: *The sample size was limited to three regions, which may not represent national trends.*
- **Exploration of unexpected results**
  - Address any findings that were not anticipated and offer plausible explanations.
  - Example: *The unexpected decline in adoption among middle-income groups may be due to higher upfront costs despite subsidies.*
- **Implications for theory and practice**
  - Discuss how the findings contribute to theoretical frameworks or practical solutions.
  - Example: *This study supports the diffusion of innovation theory, demonstrating that early adopters influence wider community adoption rates.*
- **Recommendations for future research**
  - Suggest areas where further investigation is needed to address unresolved questions or expand on your findings.
  - Example: *Future research should explore the long-term sustainability of government subsidies in driving adoption rates.*

Purpose of the discussion section

- To contextualize the results within the broader scope of the research field.
- To demonstrate the relevance and value of your findings.
- To provide a platform for advancing knowledge and informing future studies.

Formatting tips

- Begin with a summary of the key findings and their relevance.
- Use subheadings to organize themes such as implications, limitations, and future directions.

- Avoid repeating the results verbatim; focus on interpretation and significance.

The discussion section transforms raw data into meaningful insights, bridging the gap between research objectives and conclusions. It provides the rationale behind the findings and establishes their place within the larger scientific conversation.

2.8.7 Conclusion

The conclusion section encapsulates the essence of your research by summarizing the key findings, discussing their significance, and providing a roadmap for future work. It offers closure to the study while emphasizing its contributions to the field.

Key components of the conclusion section

- **Summary of key findings**
  - Recap the most important results of the study.
  - Example: *This research demonstrates that targeted government subsidies significantly increase the adoption of renewable energy technologies in urban areas.*
- **Significance of the study**
  - Highlight the broader implications of the findings for theory, practice, or policy.
  - Example: *The findings underline the need for tailored financial incentives to accelerate sustainable energy transitions globally.*
- **Alignment with objectives**
  - Confirm how the study met its research objectives.
  - Example: *By exploring regional adoption patterns, the study successfully addressed its goal of identifying barriers and drivers to renewable energy uptake.*
- **Limitations of the study**
  - Acknowledge any constraints or challenges faced during the research.
  - Example: *The study's scope was limited to urban regions, and future research should extend to rural areas for a more comprehensive understanding.*
- **Recommendations for future research**
  - Suggest areas for further investigation to build upon the study's findings.
  - Example: *Future studies could examine the long-term effects of policy changes on renewable energy adoption.*
- **Closing statement**
  - End with a strong, reflective statement that underscores the importance of the research.

- Example: *This study contributes valuable insights to the growing body of knowledge on sustainable energy practices, paving the way for impactful policy and innovation.*

Purpose of the conclusion section

- To synthesize the research into a cohesive narrative.
- To provide clarity on the study's contributions and implications.
- To inspire future research by identifying gaps and opportunities.

Formatting tips

- Avoid introducing new data or arguments in the conclusion.
- Maintain a concise and focused tone.
- Ensure that the conclusion directly ties back to the objectives stated in the introduction.

The conclusion is the final impression of your research. By summarizing findings and offering meaningful insights, it ensures your work resonates with readers and contributes constructively to the academic and professional discourse.

2.8.8 References

Key components of a references section

- **Comprehensive list of cited sources**
  - Include all works referenced in your report, such as books, journal articles, conference papers, websites, and other relevant materials.
- **Consistent citation style**
  - Follow a recognized citation style (e.g., APA, MLA, Chicago, and IEEE) based on the requirements of your institution or publisher.
  - Example (APA style):
    - Journal Article:
      *Smith, J. A., & Brown, L. T. (*2020*). The impact of renewable energy policies on urban development. Journal of Energy Studies, 35(4),* 245–260.
    - Book:
      *Johnson, M. K. (*2019*). Sustainable Practices in Urban Design. New York: Green World Publishing.*
- **Alphabetical order**
  - Arrange entries alphabetically by the last name of the first author.
- **Complete and accurate details**
  - Provide sufficient details for each source to ensure traceability.

Common citation formats

- **APA (American Psychological Association)**
  - Used in social sciences.
  - Example:
    *Brown, P., & Green, R. (*2018*). Renewable energy adoption. Energy Policy Journal, 12(3)*, 123–135.
- **MLA (Modern Language Association)**
  - Used in humanities.
  - Example:
    Smith, John. "Renewable Energy and Urban Policy." Journal of Urban Studies, vol. 35, no. 4, 2020, *pp.* 245–260.
- **Chicago Manual of Style**
  - Used in history and some sciences.
  - Example:
    Smith, John, and Linda Brown. "The Role of Policy in Energy Transition." Energy Research Journal 25, no. 4 (2020): 245–260.
- **IEEE (Institute of Electrical and Electronics Engineers)**
  - Used in engineering and technology fields.
  - Example:
    [1] J. Smith and L. Brown, "The impact of renewable energy policies on urban development," Journal of Energy Studies, vol. 35, no. 4, pp. 245–260, 2020.

Purpose of the references section

- Acknowledges the intellectual contributions of others.
- Prevents plagiarism by providing proper credit.
- Enhances credibility and traceability of your research. Tips for an effective references section
- Use reference management tools like Zotero, EndNote, or Mendeley to maintain consistency.
- Double-check all entries for accuracy and completeness.
- Ensure all in-text citations have corresponding references and vice versa.

The references section is the backbone of scholarly communication, ensuring transparency and fostering trust in your research findings.

2.9 DEVELOPING A RESEARCH PROPOSAL

A research proposal serves as a blueprint for your study, outlining its purpose, approach, and potential impact [55]. It provides a clear framework for your research while persuading reviewers of its relevance and feasibility. Include the following sections:

2.9.1 Title

The title of a research proposal should succinctly convey the essence of the study. It must be specific, clear, and engaging, offering a snapshot of the research focus. A well-crafted title should:

- Reflect on the primary topic or problem being investigated.
- Highlight the scope or key aspect of the study.
- Use precise and professional language to appeal to the intended audience.

Example

"Analyzing the Impact of Renewable Energy Policies on Urban Sustainability in Emerging Economies."

A strong title sets the tone for the proposal and captures the interest of reviewers, encouraging them to explore the study in detail.

2.9.2 Introduction

The introduction of a research proposal establishes the foundation for the study. It sets the context, defines the problem, and underscores the importance of addressing it. A well-structured introduction should include:

- **Background information:** Provide an overview of the topic, highlighting its relevance and significance.
- **Problem statement:** Clearly articulate the specific issue or gap that the research aims to address.
- **Research context:** Situate the problem within the broader field of study or industry, showcasing its importance to stakeholders.
- **Rationale:** Explain why this research is necessary and how it contributes to existing knowledge or solves a pressing issue.
- **Objectives overview:** Briefly introduce the primary objectives of the study to guide the reader towards the research goals.

Example:

"Urban sustainability is a critical challenge in emerging economies, where rapid urbanization strains resources and infrastructure. Renewable energy policies play a pivotal role in mitigating environmental impacts and fostering sustainable development. However, the lack of empirical studies assessing these policies' effectiveness presents a significant gap in the literature. This research seeks to analyze the influence of renewable energy policies on urban sustainability, providing actionable insights for policymakers and urban planners."

A compelling introduction captivates the audience, making a strong case for the proposed research.

2.9.3 Literature review

The literature review section of a research proposal provides an overview of existing research relevant to the proposed study. It identifies gaps, contextualizes the research problem, and demonstrates the importance of the study within the field. A well-constructed literature review includes:

- **Survey of existing research:** Summarize key studies, theories, and findings related to the research topic.
- **Identification of gaps:** Highlight areas where current research is lacking or where conflicting results exist.
- **Relevance to proposed study:** Explain how the reviewed literature informs or justifies the need for your research.
- **Theoretical framework:** Describe any theoretical models or frameworks that underpin the proposed study.
- **Critical analysis:** Evaluate the strengths and weaknesses of prior studies to build a case for your research.

Structure of the literature review

- **Introduction:** Briefly describe the scope of the review and its relevance to your research.
- **Thematic or chronological organization:** Organize the literature based on themes, methodologies, or historical progression.
- **Synthesis and Connection:** Relate the reviewed studies to the research objectives and questions.

Example:

"Several studies have examined the role of renewable energy policies in urban sustainability. Smith et al. (2020) explored policy frameworks in developed nations, finding a significant reduction in carbon emissions. Conversely, Gupta (2019) identified challenges in policy implementation in emerging economies due to infrastructural limitations. However, there is a paucity of comparative analyses across diverse urban contexts, particularly in low- and middle-income countries. This study seeks to address this gap by evaluating renewable energy policies in such settings, thereby contributing to the global discourse on sustainable urbanization."

A robust literature review demonstrates the researcher's understanding of the field, highlights the significance of the study, and establishes a foundation for the research methodology.

2.9.4 Objectives

The objectives section of a research proposal outlines the specific goals the study aims to achieve. Clearly defined objectives ensure that the research remains focused and aligned with its purpose. Objectives should be precise, measurable, and achievable within the given scope and timeline of the study.

Key features of well-defined objectives

- **Specificity:** Each objective should clearly state what is to be accomplished.
- **Measurability:** Define metrics or criteria to evaluate success.
- **Relevance:** Ensure alignment with the research problem and overall purpose.
- **Feasibility:** Consider resource availability and time constraints.
- **Time-Bound:** Set a realistic timeframe for achieving the objectives.

Structure of the objectives section

- **Primary objective:** The main goal of the research.
- **Secondary objectives:** Additional goals that support the primary objective or explore related aspects of the study.

Example objectives

1. To evaluate the impact of renewable energy policies on urban sustainability in low- and middle-income countries.
2. To identify the barriers to effective implementation of renewable energy initiatives in emerging economies.
3. To develop a comparative framework for analyzing policy effectiveness across diverse urban contexts.
4. To recommend actionable strategies for enhancing renewable energy adoption in under-resourced settings.

Purpose of objectives

- Provide a roadmap for the research process.
- Guide the selection of methodologies and data collection techniques.
- Facilitate the evaluation of research outcomes against predefined goals.

Clearly articulated objectives not only clarify the study's intent but also strengthen the proposal by demonstrating its direction and feasibility.

2.9.5 Methodology

The methodology section of a research proposal outlines the approach, techniques, and tools that will be used to conduct the study [15, 25].

This section ensures that the research process is systematic, reproducible, and aligned with the objectives. It also demonstrates the feasibility and rigour of the proposed methods to reviewers.

Key components of the methodology section

- **Research design**
  - Specify whether the study is experimental, observational, descriptive, or analytical.
  - Explain the rationale behind selecting the design.
- **Data collection methods**
  - Describe the techniques for gathering data, such as surveys, interviews, experiments, or archival research.
  - Detail the tools and instruments (e.g., questionnaires, sensors, and software) to be used.
- **Population and sampling**
  - Define the target population for the study.
  - Explain the sampling method (e.g., random sampling and stratified sampling) and justify its selection.
  - Mention the sample size and its determination process.
- **Variables and measurements**
  - Identify independent, dependent, and control variables.
  - Explain how each variable will be measured or quantified.
- **Data analysis techniques**
  - Specify statistical or qualitative methods for analyzing data (e.g., regression analysis and thematic coding).
  - Mention any software or tools (e.g., SPSS, NVivo, and MATLAB) to be employed.
- **Ethical considerations**
  - Outline measures to ensure ethical compliance, such as informed consent, anonymity, and data security.
  - Highlight adherence to relevant guidelines and standards.
- **Limitations and delimitations:**
  - Acknowledge potential constraints (e.g., time, resources, and access to data).
  - Define the boundaries of the study to manage expectations.

Example methodology overview

This study adopts a mixed-methods approach, combining qualitative interviews with quantitative surveys. Data will be collected from 200 participants selected through stratified random sampling. Surveys will assess energy consumption patterns, while interviews will explore attitudes towards renewable energy policies. Data will be analyzed using thematic analysis for qualitative data and regression modelling for quantitative data. Ethical clearance will be obtained prior to data collection.

Purpose of the methodology section

- Provide a transparent and replicable plan for conducting the research.
- Justify the appropriateness of methods for achieving the objectives.
- Assure reviewers of the study's scientific rigour and ethical soundness.

A well-detailed methodology enhances the credibility of the research proposal and lays the groundwork for producing reliable and impactful findings.

2.9.6 Timeline

The timeline section of a research proposal provides a structured schedule for completing the study. It ensures that the research is feasible within the given time frame and helps reviewers understand the sequence and duration of tasks. A clear timeline demonstrates the researcher's planning and organizational skills.

Key components of the timeline section

- **Phases of the study**
  - Break down the research process into distinct phases (e.g., planning, data collection, and analysis).
  - Assign specific time periods for each phase.
- **Milestones and deliverables**
  - Identify key milestones, such as literature review completion, data collection, and submission of interim reports.
  - Include deliverables, such as draft chapters, presentations, or publications.
- **Time allocation**
  - Provide an estimate of the time required for each activity (e.g., weeks or months).
  - Ensure that overlapping tasks are accounted for where feasible.
- **Presentation format**
  - Use a Gantt chart or tabular format to visually represent the timeline.

Example timeline (6-month study) (Table 2.1):

Purpose of the timeline section

- Demonstrates the feasibility of completing the research within the proposed duration.
- Helps track progress and ensure timely completion of tasks.
- Facilitates coordination among collaborators and stakeholders.

A detailed and realistic timeline reassures reviewers that the researcher is well-prepared to execute the study efficiently and effectively.

Table 2.1 Example of timeline (6-month study)

Phase	*Activities*	*Duration*	*Milestones*
Phase 1: Planning	Define the research problem and objectives	2 weeks	Approved research proposal
Phase 2: Literature Review	Review and synthesize existing studies	4 weeks	Comprehensive literature review
Phase 3: Data Collection	Design instruments, collect data	8 weeks	Validated dataset
Phase 4: Data Analysis	Analyze and interpret findings	6 weeks	Statistical and thematic results
Phase 5: Reporting	Write reports, draft publications	6 weeks	Draft report and papers
Phase 6: Finalization	Revise and submit final documents	2 weeks	Final submission

2.9.7 Budget

The budget section of a research proposal outlines the estimated financial resources required to conduct the study. It provides transparency and demonstrates that the researcher has carefully considered the costs associated with the project. A well-prepared budget helps reviewers assess the feasibility and practicality of the research.

Key components of the budget section

1. **Personnel costs**
   - Salaries or stipends for researchers, assistants, or consultants.
   - Include time commitments and hourly rates if applicable.
2. **Equipment and supplies**
   - Costs for purchasing or renting specialized equipment, tools, or software.
   - List essential consumables like laboratory materials, stationery, or data storage devices.
3. **Travel expenses**
   - Transportation, lodging, and meals for fieldwork, conferences, or meetings.
   - Include local travel for data collection or stakeholder consultations.
4. **Data collection costs**
   - Survey tools, printing questionnaires, or compensating participants.
   - Costs for accessing databases or purchasing datasets.
5. **Publication and dissemination**
   - Fees for publishing articles, preparing reports, or designing presentations.
   - Printing costs for hard copies of deliverables.

Table 2.2 Example of budget table

Category	*Description*	*Estimated Cost (USD)*
Personnel	Research assistant (6 months)	3,000
Equipment	Statistical software license	500
Travel	Field visits (transportation and lodging)	1,200
Data Collection	Printing surveys and participant compensation	800
Publication	Journal article submission fee	1,000
Miscellaneous	Contingency (10% of total)	650
Total		7,150

6. **Miscellaneous expenses**
   - Unexpected or indirect costs, such as utilities, insurance, or administrative fees.
   - A contingency fund (usually 5–10% of the total budget).

Example budget table (Table 2.2)

Purpose of the budget section

- Ensures adequate planning and allocation of resources.
- Helps reviewers evaluate the cost-effectiveness of the study.
- Demonstrates the researcher's understanding of project requirements.

A clear and detailed budget, supported by justifications, enhances the credibility of the proposal and reassures funding agencies or institutions of its viability.

2.9.8 Expected outcomes

The "Expected Outcomes" section of a research proposal outlines the anticipated results and their potential significance. It highlights the contributions the research will make to the field and society, ensuring that reviewers understand its value and impact.

Key components of expected outcomes

- **Research findings**
  - Specific insights or data the study aims to uncover.
  - Examples: Patterns, relationships, or evidence supporting or refuting hypotheses.
- **Contribution to knowledge**
  - How the research will fill gaps identified in the literature review.
  - Theoretical advancements or practical applications resulting from the study.

- **Practical implications**
  - Real-world benefits or applications of the research.
  - Examples: Improved technologies, better policies, or enhanced understanding of critical issues.
- **Future research directions**
  - How the outcomes will pave the way for subsequent studies.
  - Identification of new questions or challenges arising from the findings.
- **Impact on stakeholders**
  - Specific benefits for the target audience, policymakers, industry, or community.
  - Examples: Improved processes, solutions to pressing problems, or empowerment of specific groups.

Purpose of the expected outcomes section

- Demonstrates the relevance and significance of the research.
- Provides clarity on what the study aims to achieve.
- Aligns the research objectives with tangible results, ensuring feasibility and focus.

Example of expected outcomes

- **For a study on renewable energy adoption**
  - Identification of key factors influencing adoption rates.
  - Development of a framework for policymakers to promote renewable energy.
- **For research on improving educational tools**
  - Evidence of increased student engagement and learning outcomes.
  - Recommendations for implementing the tools across institutions.

By clearly defining the expected outcomes, the proposal not only establishes the importance of the research but also builds confidence in its ability to deliver meaningful results.

2.10 PRESENTATION AND ASSESSMENT BY A REVIEW COMMITTEE

A presentation and assessment by a review committee typically involves a structured process where an individual or team presents their work, research, or project to a panel of experts or stakeholders [56]. The committee then evaluates the presentation based on specific criteria. Here's an outline of the process:

- **Preparation**
  - **Presentation content:** Prepare clear and concise slides or materials that summarize the key aspects of the project, research, or work.

  - **Rehearsal:** Practice delivering the presentation to ensure clarity and confidence.
  - **Documentation:** Ensure all relevant documentation, data, and reports are ready for review.
- **Presentation**
  - **Introduction:** Briefly introduce yourself, your background, and the purpose of your work.
  - **Objectives:** Outline the goals and objectives of your project or research.
  - **Methodology:** Present the approach, techniques, or methods used.
  - **Results:** Share the outcomes, findings, or solutions derived from your work.
  - **Conclusions:** Summarize the main takeaways, significance, and future directions.
- **Assessment**
  - **Clarity and structure:** The committee evaluates the organization, clarity, and effectiveness of the presentation.
  - **Content Accuracy:** Reviewers assess the technical or academic rigour of the content presented.
  - **Innovation:** The novelty or creativity of the work is evaluated.
  - **Impact:** The potential implications or applications of the work are considered.
  - **Engagement:** The ability of the presenter to engage and respond to questions or feedback from the committee.
- **Feedback**
  - After the presentation, the committee provides feedback, which may include strengths, areas for improvement, and suggestions for further development.
  - Recommendations for future work or adjustments are often given.
- **Decision**
  - Based on the assessment, the committee may make decisions regarding the next steps, approval, funding, or publication of the work.

This type of review process is commonly used in academic, research, business, and technical fields to ensure the quality and viability of projects or ideas.

2.11 SUMMARY

This chapter covers essential concepts and best practices for conducting effective literature studies, ensuring high-quality research and writing. It emphasizes the importance of clearly defining objectives, utilizing diverse sources, and employing advanced search techniques to gather comprehensive and relevant information. Critical evaluation of sources is crucial, as it allows researchers to assess the validity and reliability of the literature. The chapter outlines methods for organizing findings logically, including thematic grouping, chronological order, and cause-and-effect relationships, as well as comparative and hierarchical analysis. Highlighting key findings and documenting references properly is fundamental to maintaining clarity and transparency in research.

In-depth analysis of literature is central to understanding the context, trends, and gaps in a particular field. Steps in literature analysis include identifying core themes, evaluating methodologies, assessing relationships, and reviewing outcomes. Synthesizing findings from various sources strengthens research arguments and helps inform future research design. Literature analysis plays a pivotal role in establishing the research context, identifying trends, and supporting the overall validity of research claims.

The chapter also addresses plagiarism in research, detailing the different types, including direct, self, mosaic, and accidental plagiarism, and their consequences such as reputational damage, legal issues, and academic penalties. Preventing plagiarism is vital, with strategies such as proper citation, paraphrasing, and using plagiarism detection tools. The importance of research integrity is reinforced through a focus on ethical practices like maintaining confidentiality, informed consent, and non-maleficence.

Technical writing skills are essential for producing clear, consistent, and accurate reports, with a focus on audience understanding, logical flow, and proper documentation. The chapter outlines the steps for writing a report, including crafting the title page, abstract, introduction, methodology, results, discussion, conclusion, and references. Furthermore, it provides a framework for developing a research proposal, highlighting key components such as the title, objectives, methodology, timeline, budget, and expected outcomes. Finally, the chapter covers the presentation and assessment process by a review committee, stressing the importance of following organizational standards and guidelines while effectively addressing feedback and questions.

REVIEW QUESTIONS

1. What are the key strategies for conducting effective literature studies?
2. How can you ensure your objectives are clearly defined in a literature review?
3. Why is it important to use diverse sources when reviewing literature?
4. What are some advanced search techniques you can use to gather relevant literature?
5. How can you critically evaluate the sources in your literature review?
6. What is the importance of logically organizing findings in a literature review?

7. How does thematic grouping improve the structure of a literature review?
8. What are the advantages of presenting findings in chronological order?
9. How can cause-and-effect relationships be highlighted in a literature review?
10. What is comparative analysis, and how does it contribute to literature review?
11. How does hierarchical structuring enhance the readability of a literature review?
12. Why is it important to highlight key findings in a literature review?
13. What are the steps involved in literature analysis?
14. How do you identify core themes in literature?
15. What should be considered when evaluating the methodologies used in previous research?
16. Why is it important to highlight gaps in existing literature?
17. How do you assess relationships between different pieces of literature?
18. What should you look for when reviewing the outcomes of a study in your literature review?
19. How can you critique the validity of sources used in your literature analysis?
20 What is the role of synthesizing findings in a literature review?
21. How does literature analysis inform research design?
22. In what ways does literature analysis establish the context for new research?
23. How can literature analysis help identify trends in a particular field?
24. Why is literature analysis important for strengthening research arguments?
25. What are the different types of plagiarism in research?
26. How does direct plagiarism differ from self-plagiarism?
27. What is mosaic plagiarism, and how does it affect academic integrity?
28. What are the consequences of plagiarism in academic and professional settings?
29. How can plagiarism hinder scientific progress?
30. What strategies can be used to prevent plagiarism in research?
31. What are the key principles of research ethics?
32. Why is integrity an important principle in research ethics?
33. How does confidentiality protect research participants?
34. What is informed consent, and why is it crucial in research ethics?
35. How does the principle of non-maleficence apply to research?
36. Why is respect for participants a core ethical consideration in research?
37. How does accountability play a role in ethical research practices?
38. Why is justice important in research ethics?
39. How do research ethics protect participants in a study?
40. In what ways do research ethics promote trust in scientific findings?
41. How do research ethics enhance the credibility of a study?

42. What are the essential skills needed for effective technical writing?
43. Why is understanding the audience crucial in technical writing?
44. How does writing for clarity improve technical documents?
45. What role do grammar and punctuation play in technical writing?
46. Why is consistency important in technical writing?
47. How can logical flow be achieved in technical writing?
48. Why is accuracy in information important in technical writing?
49. How do you decide when to use tables and figures in technical writing?
50. What is the purpose of technical writing, and how do different types of technical writing serve this purpose?
51. What is the purpose of including a title page in a report?
52. What should be included in the abstract of a report?
53. How do you structure the introduction section of a report?
54. Why is the methodology section important in a report?
55. What is the purpose of the results section in a report?
56. How do you write a discussion section in a report?
57. What should be covered in the conclusion of a report?
58. Why is it essential to include references in a report?
59. What should be included in the title of a research proposal?
60. Why is the introduction crucial in a research proposal?
61. What is the role of the literature review in a research proposal?
62. How do you define clear objectives in a research proposal?
63. What should the methodology section in a research proposal cover?
64. Why is a timeline important in a research proposal?
65. How does a budget support the development of a research proposal?
66. What should be included in the expected outcomes of a research proposal?
67. What are the key elements of a successful presentation to a review committee?
68. How can you effectively address questions and feedback from a review committee?
69. What are the criteria typically used by a review committee to assess a project or research?
70. Why is it important to follow organizational standards and guidelines during a review presentation?

Chapter 3

Key skills for effective paper writing

3.1 INTRODUCTION

Writing a research paper is a fundamental aspect of academic and professional work, serving as a platform to share findings, insights, and innovations with the broader community. However, the process of crafting a well-structured and impactful paper requires more than just presenting data; it demands mastery of specific skills to ensure clarity, coherence, and credibility.

This chapter, "Key Skills for Effective Paper Writing," delves into the essential competencies required to create compelling research papers. From choosing a concise yet descriptive title to presenting findings and discussing their implications, each section of a paper demands a tailored approach [57]. By understanding and applying these skills, researchers can enhance the quality of their work, effectively engage their audience, and meet the rigorous standards of academic and professional writing.

Whether you are a novice writer aiming to publish your first paper or an experienced researcher seeking to refine your craft, this chapter provides a comprehensive guide to mastering the art of paper writing. Through practical insights and actionable techniques, it equips you with the tools needed to communicate your research effectively and contribute meaningfully to your field.

3.2 SECTIONS OF A PAPER

A research paper is a structured document that communicates findings in a systematic and logical manner [58]. Each section has a distinct role, contributing to the overall clarity and impact of the paper. Below is an in-depth look at the key sections:

DOI: 10.1201/9781003485827-3

Abstract

The abstract is a concise summary of the research, typically 150–250 words. It highlights:

- **Objectives**: What the study aims to achieve.
- **Methods**: How the study was conducted.
- **Results**: Key findings.
- **Conclusions**: The significance of the study.

A strong abstract captures the essence of the paper, enabling readers to quickly assess its relevance.

Review of the literature

This section synthesizes existing research on the topic, emphasizing:

- Current knowledge and understanding.
- Gaps in the literature that the study addresses.
- Theoretical or practical frameworks underpinning the research.

A well-written review of the literature demonstrates the researcher's depth of knowledge and positions the study within the field.

Methods

The methods section provides a detailed account of how the research was conducted, including:

- Study design (e.g., experimental and observational).
- Data collection techniques.
- Tools and instruments used.
- Statistical or analytical methods.

This section ensures reproducibility and allows others to evaluate the validity of the study.

Results

The results section presents the findings of the study, often supported by:

- Tables
- Figures
- Graphs

It focuses on presenting data objectively without interpretation, ensuring clarity and precision.

Discussion

The discussion interprets the results, connecting them to the research objectives and existing literature. It includes:

- Implications of the findings.
- Comparisons with prior studies.
- Limitations of the research.

This section explores the broader significance of the study and suggests future directions.

Conclusions

The conclusion summarizes the study's key findings and their implications. It reinforces the importance of the research and often suggests practical applications or areas for further investigation.

The final check

Before submission, the paper undergoes a final review to ensure:

- Accuracy of data and references.
- Consistency in formatting and style.
- Clarity and coherence of writing.

This step ensures the paper meets academic and professional standards, increasing its likelihood of acceptance and impact.

3.3 KEY SKILLS ARE NEEDED WHEN WRITING AN INTRODUCTION

The title of a research paper is the first element readers encounter, and it significantly impacts their decision to read further [59]. Crafting an effective title requires a balance of clarity, conciseness, and relevance. Here are the key skills essential for writing an impactful title:

3.3.1 Clarity

Clarity is vital when crafting the introduction of a research paper, as it sets the stage for the reader and ensures the purpose of the study is well understood. Below are the key aspects of achieving clarity when writing an introduction:

State the research problem clearly

- **Why it matters:** A vague or overly complex problem statement can confuse readers and obscure the focus of the study.

- **How to achieve it:** Use precise language to articulate the problem your research addresses. For example, instead of "This study looks at issues in renewable energy," write, "This study investigates the efficiency challenges in solar panel technology under varying weather conditions."

Provide context and background information

- **Why it matters:** Insufficient context can leave readers guessing about the relevance of your study.
- **How to achieve it:** Briefly introduce the background of the research area using simple and direct language. Avoid overloading this section with technical jargon or excessive detail.

Define the scope of the study

- **Why it matters:** A poorly defined scope can lead to misunderstandings about the study's focus and limitations.
- **How to achieve it:** Specify the boundaries of your research clearly, such as the population, time frame, or specific aspects of the problem being addressed.

Articulate the objectives

- **Why it matters:** Ambiguity in research objectives can make it hard for readers to grasp the purpose of the study.
- **How to achieve it:** Use concise statements to explain what the research aims to achieve. For example, "This study aims to analyze the impact of urbanization on local biodiversity."

Highlight the research significance

- **Why it matters:** Readers need to understand why your study matters to the field or society.
- **How to achieve it:** Clearly explain the potential impact or contributions of your research without exaggeration. For example, "Understanding this relationship could help policymakers create more effective conservation strategies."

Maintain logical flow

- **Why it matters:** Disorganized ideas can confuse readers and reduce the effectiveness of your introduction.
- **How to achieve it:** Arrange the introduction logically—start with general context, narrow down to the specific problem, and conclude with the study's objectives.

Avoid unnecessary jargon

- **Why it matters:** Overuse of technical terms can alienate readers unfamiliar with the subject.
- **How to achieve it:** Simplify complex ideas without compromising accuracy. For example, explain technical terms briefly when first introduced.

Write concisely

- **Why it matters:** Overly lengthy introductions can lose the reader's interest.
- **How to achieve it:** Keep sentences and paragraphs focused on conveying essential information without digression.

Pose research questions (if applicable)

- **Why it matters:** Unclear or overly broad questions can detract from the focus of the introduction.
- **How to achieve it:** Clearly articulate specific, measurable, and relevant research questions that align with your objectives.

By focusing on clarity, the introduction becomes an engaging and informative gateway to the research, guiding readers effortlessly into the study's purpose and significance.

3.3.2 Conciseness

Conciseness is a crucial skill when writing the introduction of a research paper. It ensures the section is focused, engaging, and informative without overwhelming the reader with unnecessary details. Here's how conciseness can be achieved:

Avoid redundancy

- **Why it matters:** Repetition can make the introduction lengthy and tedious.
- **How to achieve it:** Present each point only once, ensuring that sentences are clear and distinct. For example, instead of "This research is important and valuable to the field because it is significant," write, "This research is significant for advancing the field."

Prioritize key information

- **Why it matters:** Including too much background or tangential details can detract from the main message.
- **How to achieve it:** Focus on the essentials—state the research problem, provide necessary context, and highlight objectives. Leave detailed explanations for later sections.

Use precise language

- **Why it matters:** Wordiness can obscure the main points and confuse readers.
- **How to achieve it:** Replace lengthy phrases with concise alternatives. For instance, use "to analyze" instead of "in order to perform an analysis of."

Avoid overloading with background information

- **Why it matters:** Over-explaining the context can make the introduction disproportionately long.
- **How to achieve it:** Limit background details to what is directly relevant to the research problem. Reserve additional context for the literature review.

Write compact sentences

- **Why it matters:** Long, complex sentences can reduce readability and impact.
- **How to achieve it:** Break up long sentences into shorter, more direct ones while maintaining logical flow.

Stick to the word count

- **Why it matters:** Exceeding the word count can reduce the overall quality of the paper and strain the reader's attention.
- **How to achieve it:** Regularly revise and edit to eliminate superfluous words or ideas.

Use a logical structure

- **Why it matters:** Rambling introductions can lose the reader's focus.
- **How to achieve it:** Follow a structured approach—start with the broad context, narrow to the research problem, and end with the objectives or significance.

Avoid irrelevant information

- **Why it matters:** Including unrelated details dilutes the purpose of the introduction.
- **How to achieve it:** Keep every sentence aligned with the research objectives and problem.

Use bullet points (if appropriate)

- **Why it matters:** For grant proposals or technical reports, a concise list format can convey key ideas quickly.
- **How to achieve it:** Summarize objectives or questions in a bulleted or numbered list to save space and improve clarity.

By mastering conciseness, the introduction remains impactful and reader-friendly, ensuring the paper's purpose is communicated efficiently without unnecessary elaboration.

3.3.3 Relevance

Relevance is a critical skill in crafting an introduction that resonates with the audience and sets the stage for the research. It ensures that the information presented is meaningful, directly connected to the research problem, and aligned with the paper's objectives.

Align with the research problem

- **Why it matters:** A relevant introduction connects background information and context directly to the research problem.
- **How to achieve it:** Clearly state how the problem relates to existing gaps or challenges in the field.

Tailor to the target audience

- **Why it matters:** Different audiences, such as experts, policymakers, or students, require varying levels of detail and emphasis.
- **How to achieve it:** Use terminology, examples, and references that align with the audience's knowledge and interests.

Highlight the significance of the study

- **Why it matters:** The introduction must demonstrate why the research is important and how it contributes to the field.
- **How to Achieve It:** Explicitly connect the study to broader scientific, societal, or industry needs.

Include pertinent background information

- **Why it matters:** Background details should support the research question rather than distract from it.
- **How to achieve it:** Focus on the most relevant theories, prior studies, or historical context that lead to the current research.

Avoid tangential information

- **Why it matters:** Irrelevant details can confuse readers and detract from the main purpose.
- **How to achieve it:** Ensure every sentence contributes directly to understanding the research problem or objectives.

Address current trends and needs

- **Why it matters:** A relevant introduction ties the study to ongoing developments or emerging issues in the field.
- **How to achieve it:** Mention recent advancements, challenges, or controversies that your research addresses.

Cite appropriate sources

- **Why it matters:** References to outdated or unrelated studies reduce the credibility and relevance of the introduction.
- **How to achieve it:** Use recent and authoritative sources that support the research problem and objectives.

Define the scope clearly

- **Why it matters:** A focused scope ensures the introduction remains relevant and concise.
- **How to achieve it:** Specify what the study will and will not address, avoiding overly broad claims.

Connect to objectives and outcomes

- **Why it matters:** Readers need to see a clear link between the introduction and the study's goals.
- **How to achieve it:** End the introduction by tying the background and problem statement to the research objectives or hypotheses.

By emphasizing relevance, the introduction ensures that the reader understands the importance of the research and remains engaged with its purpose, setting a strong foundation for the rest of the paper.

3.3.4 Use of keywords

Keywords are essential for enhancing the discoverability, clarity, and impact of a research paper. Proper use of keywords in the introduction ensures that the research aligns with its audience and meets academic and professional standards.

Highlight core concepts

- **Why it matters:** Keywords convey the main ideas and themes of the research, guiding readers to the essence of the study.
- **How to achieve it:** Incorporate terms that reflect the research problem, objectives, and methodology.

Improve searchability

- **Why it matters:** Keywords optimize the paper for search engines and databases, increasing visibility.
- **How to achieve it:** Use specific and widely recognized terms related to the topic, avoiding overly general or niche jargon.

Align with field-specific terminology

- **Why it matters:** Consistency with established vocabulary ensures the paper appeals to its intended academic or professional audience.
- **How to achieve it:** Include terms commonly used in the field or discipline of the research.

Maintain relevance

- **Why it matters:** Irrelevant or overly broad keywords can mislead readers or reduce the paper's visibility in targeted searches.
- **How to achieve it:** Select keywords that accurately reflect the content and scope of the study.

Support literature context

- **Why it matters:** Keywords help connect the research to existing studies and debates in the field.
- **How to achieve it:** Choose terms that overlap with those in cited works, ensuring continuity in the scholarly conversation.

Integrate naturally into the text

- **Why it matters:** Keywords that feel forced or repetitive can disrupt the flow of the introduction.
- **How to achieve it:** Use keywords naturally within sentences, ensuring they fit the context.

Use keywords to frame the problem

- **Why it matters:** Well-chosen keywords provide a concise snapshot of the research problem.
- **How to achieve it:** Introduce keywords early in the introduction to set the tone and context.

Facilitate cross-disciplinary reach

- **Why it matters:** Keywords can help bridge gaps between disciplines, attracting a broader audience.
- **How to achieve it:** Include terms that are relevant across multiple fields, if applicable.

Balance specificity and breadth

- **Why it matters:** Keywords should be precise enough to target the right audience but broad enough to ensure visibility.
- **How to achieve it:** Use a mix of specific terms (e.g., "finite element analysis") and general ones (e.g., "mechanical engineering").

Revisit and refine

- **Why it matters:** Keywords should evolve as the paper develops to ensure alignment with the final content.
- **How to achieve it:** Review keywords after completing the paper to confirm their accuracy and relevance.

By strategically using keywords in the introduction, authors can enhance the paper's accessibility, relevance, and integration into academic and professional discourse.

3.3.5 Avoiding jargon

Jargon refers to specialized terms or language that may be difficult for readers outside a specific field to understand. While technical terminology is often necessary, excessive or inappropriate use of jargon can alienate readers and obscure the paper's message. Avoiding jargon ensures clarity, accessibility, and broader appeal in your introduction.

Recognize your audience

- **Why it matters:** Not all readers will have the same level of expertise as the author.
- **How to achieve it:** Tailor your language to match the knowledge level of your target audience, ensuring they can follow the content without difficulty.

Use plain language

- **Why it matters:** Simplifying complex ideas improves understanding and retention.
- **How to achieve it:** Replace technical terms with common language where possible, and provide explanations for necessary technical terms.

Explain necessary terms

- **Why it matters:** Some jargon may be unavoidable in specialized research, but it should not confuse the reader.
- **How to achieve it:** Define any technical terms or acronyms the first time they appear.

Focus on clarity

- **Why it matters:** Jargon can obscure the main message, making it harder for readers to grasp the research's purpose.
- **How to achieve it:** Write concise, straightforward sentences that convey key points without unnecessary complexity.

Avoid overloading sentences

- **Why it matters:** Dense sentences filled with jargon are hard to read and understand.
- **How to achieve it:** Break complex ideas into shorter, simpler sentences, avoiding clusters of technical terms.

Use analogies or examples

- **Why it matters:** Analogies and examples can help explain complex ideas in a relatable way.
- **How to achieve it:** Use comparisons or real-world examples to clarify technical concepts for a broader audience.

Avoid buzzwords

- **Why it matters:** Buzzwords can be vague and detract from the credibility of the introduction.
- **How to achieve it:** Use precise language instead of trendy or overused terms that lack substantive meaning.

Test for accessibility

- **Why it matters:** If the introduction is inaccessible, readers may lose interest or fail to understand its significance.
- **How to achieve it:** Have someone outside your field read the introduction and provide feedback on clarity and comprehension.

Strike a balance

- **Why it matters:** The introduction should be accessible without oversimplifying key concepts.
- **How to achieve it:** Maintain a professional tone while ensuring that technical details are presented clearly and effectively.

Revisit during editing

- **Why it matters:** Jargon often creeps into drafts unintentionally.
- **How to achieve it:** During the revision process, identify and replace jargon with clearer language wherever possible.

By minimizing jargon, authors can make their introductions more inclusive and impactful, ensuring that readers from various backgrounds can engage with and understand the research.

3.3.6 Specificity

Specificity refers to the clarity and precision with which ideas are presented, ensuring the introduction is focused, informative, and aligned with the research's objectives. Being specific helps readers quickly understand the scope, significance, and direction of the study.

Define the research problem clearly

- **Why it matters:** A vague or broad problem can confuse readers about the study's purpose.
- **How to achieve it:** State the research problem in precise terms, avoiding ambiguity or overly general statements.

Narrow the focus

- **Why it matters:** A specific introduction avoids overwhelming the reader with unrelated or excessive background information.
- **How to achieve it:** Highlight only the most relevant context and omit details that do not directly support the research objectives.

Use quantifiable details

- **Why it matters:** Generalizations can make the introduction seem unsubstantiated.
- **How to achieve it:** Include measurable data, examples, or statistics that support the research's importance.

Provide a targeted background

- **Why it matters:** An overly broad background may distract readers from the core research focus.
- **How to achieve it:** Discuss only the specific aspects of the field that directly relate to the research problem.

Articulate clear objectives

- **Why it matters:** Specific objectives give readers a clear understanding of what the study aims to achieve.
- **How to achieve it:** Use concise language to outline the study's goals, avoiding vague terms like "explore" or "examine" without context.

Avoid overgeneralized claims

- **Why it matters:** Sweeping statements can undermine credibility and mislead readers about the study's scope.
- **How to achieve it:** Focus on specific claims backed by evidence or logical reasoning.

Tailor examples to the research context

- **Why it matters:** Generic examples may fail to illustrate the relevance of the study.
- **How to achieve it:** Use examples directly related to the research topic to clarify complex ideas.

Frame a precise research question

- **Why it matters:** A well-defined research question provides direction and focus.
- **How to achieve it:** Craft a research question that is narrow, feasible, and aligned with the study's goals.

Link to the study's contribution

- **Why it matters:** Specificity highlights the unique value of the research within its field.
- **How to achieve it:** Clearly state how the research addresses a gap or contributes new insights.

Revise for precision

- **Why It matters:** Initial drafts may contain broad or unclear statements.
- **How to achieve it:** Edit the introduction to remove generalizations, ensuring every sentence is focused and purposeful.

By emphasizing specificity, the introduction becomes a compelling roadmap for the paper, guiding readers through the research's relevance and objectives without ambiguity. This approach builds credibility and fosters engagement with the study.

3.3.7 Balance between creativity and formality

Achieving a balance between creativity and formality ensures that an introduction is engaging while maintaining the professional tone required in academic or technical writing [60]. This balance helps capture the reader's interest without compromising the rigour or credibility of the research.

Creativity: engaging the reader

- **Why it matters:** A creative approach can make the introduction more appealing and memorable.

- **How to achieve it:**
  - Begin with a thought-provoking question, anecdote, or striking statistic relevant to the research.
  - Use vivid language to describe the significance of the study, but avoid exaggeration or unnecessary embellishments.
  - Incorporate analogies or metaphors sparingly to simplify complex ideas.

Formality: establishing credibility

- **Why it matters:** A formal tone ensures the introduction is taken seriously by the academic or professional audience.
- **How to achieve it:**
  - Use precise, professional language and avoid colloquialisms or overly casual expressions.
  - Structure sentences logically and adhere to grammatical norms.
  - Maintain objectivity by avoiding overly emotional or subjective statements.

Balancing creativity and formality

- **Why it matters:** Striking this balance ensures the introduction is both captivating and authoritative.
- **How to achieve it:**
  - Open creatively to draw the reader in, but transition quickly into a formal discussion of the research problem.
  - Ensure creative elements support, rather than overshadow, the core objectives of the introduction.
  - Avoid excessive technical jargon while maintaining the integrity of the content.

Example of balanced writing

Creative opening

- "Imagine a world where renewable energy sources power every home and vehicle—this vision drives the urgent need for breakthroughs in battery technology."

Formal transition

- "This study investigates the efficiency of lithium-ion batteries, focusing on improving their storage capacity and reducing environmental impact."

Tips for maintaining balance

- **Review audience expectations:** Tailor the balance based on whether the audience values innovation, strict professionalism, or both.

- **Limit creativity to the opening:** Use creative techniques sparingly, primarily in the opening sentences, to set the stage.
- **Use evidence-based statements:** Anchor creative elements with facts, data, or logical reasoning to ensure credibility.
- **Seek feedback:** Have peers or mentors review the introduction to ensure it strikes the right tone.

By blending creativity with formality, writers can craft introductions that resonate with readers, engage their interests, and uphold the standards of academic or technical writing.

3.3.8 Following journal or institution guidelines

Adhering to journal or institutional guidelines is essential for ensuring that your paper meets the expectations of the target audience and avoids unnecessary revisions or rejections [61]. These guidelines often dictate the structure, tone, and content of the introduction, providing a framework that aligns with the publication's standards.

Importance of guidelines

- **Consistency:** Ensures uniformity across submissions, making it easier for reviewers and editors to assess papers.
- **Compliance:** Demonstrates respect for the journal's or institution's requirements, increasing the likelihood of acceptance.
- **Efficiency:** Reduces time spent on reformatting or revising due to non-compliance.

Key aspects of guidelines for the introduction

- **Length restrictions:** Many journals specify word limits for introductions. Stay concise and within the specified range.
- **Structure:** Guidelines may outline the sequence of elements, such as background information, problem statement, objectives, and scope.
- **Tone and style:** Some journals emphasize a formal tone, while others may allow for a slightly conversational style depending on the audience.
- **Inclusion of keywords:** Specific terms or phrases relevant to the field may be required to enhance discoverability.
- **References:** Journals may provide instructions on citing prior works directly in the introduction.

Steps to follow guidelines

- **Read the author's instructions:** Carefully review the journal's or institution's submission guidelines before drafting.

- **Analyze published examples:** Study introductions from previously published articles in the target journal to understand the expected format and style.
- **Use templates or checklists:** Many journals provide templates or checklists for authors to ensure compliance.
- **Consult editorial queries:** If guidelines are unclear, contact the journal editor for clarification.

Common requirements

- **Focus on relevance:** Clearly link the introduction to the journal's scope or research focus.
- **Avoid excessive background:** Journals often discourage overly detailed literature reviews in the introduction, reserving them for a separate section.
- **State novelty:** Emphasize the unique contribution of your research early in the introduction.
- **Follow formatting rules:** Adhere to font, spacing, and citation style specified by the guidelines.

Example of guideline adherence

- **Journal guideline:** "The introduction should be no more than 300 words, state the research problem, and include a maximum of three references."
- **Draft example:** "This study addresses the rising challenge of antibiotic resistance in clinical settings. Building on previous research (Smith et al., 2020; Jones, 2019), we explore the efficacy of novel antimicrobial peptides. The findings aim to contribute to the development of alternative treatment strategies, aligning with global health priorities."

Tips for compliance

- **Double-check before submission:** Review the introduction against the guidelines to ensure all requirements are met.
- **Seek feedback:** Share your draft with colleagues or mentors familiar with the journal's expectations.
- **Use formatting tools:** Leverage citation managers and document templates to align with specific formatting rules.

By closely following journal or institutional guidelines, writers demonstrate professionalism, improve their paper's chances of acceptance, and streamline the publication process.

3.4 SKILLS NEEDED WHEN WRITING A REVIEW OF THE LITERATURE

A literature review is the backbone of a research paper, offering context, identifying gaps, and establishing the foundation for the study [62]. Writing an effective review requires several key skills to ensure clarity, coherence, and relevance.

3.4.1 Critical thinking

Critical thinking is an essential skill for writing a compelling and meaningful review of the literature. It involves analyzing, synthesizing, and evaluating existing research to establish the foundation for your study and highlight its relevance and originality.

Importance of critical thinking in literature reviews

- **Identifying gaps:** Helps pinpoint areas where existing research is insufficient or inconsistent.
- **Evaluating credibility:** Ensures the use of reliable, high-quality sources.
- **Avoiding bias:** Encourages objective analysis rather than personal opinions or selective reporting.
- **Building arguments:** Strengthen the rationale for your study by linking findings logically.

Application of critical thinking

- **Analyze key studies:** Examine methodologies, results, and conclusions to understand their strengths and weaknesses.
- **Compare and contrast:** Highlight similarities and differences between studies to identify patterns or contradictions.
- **Assess relevance:** Focus on research that directly informs your topic or fills a gap in knowledge.
- **Synthesize information:** Integrate findings from various studies to provide a cohesive narrative.

Steps for applying critical thinking

- **Question assumptions:** Challenge the premises and conclusions of studies to uncover potential biases or limitations.
- **Evaluate methodologies:** Consider whether the research methods used in the studies are appropriate and robust.
- **Look for contradictions:** Identify conflicting evidence and explore possible reasons for discrepancies.
- **Determine significance:** Assess the impact and relevance of each study to your research question.

Example of critical thinking in practice

Without critical thinking:

"Several studies have explored the effects of exercise on mental health (Smith et al., 2020; Brown et al., 2021)."

With critical thinking:

"While Smith et al. (2020) demonstrated significant mental health benefits of moderate exercise, Brown et al. (2021) found negligible effects in older adults, possibly due to differing sample characteristics. This inconsistency highlights the need for further research into age-specific exercise interventions."

Tips for developing critical thinking

- **Ask questions:** Who conducted the study? What were their motivations? Are the results generalizable?
- **Stay objective:** Avoid letting personal biases influence your evaluation of the literature.
- **Engage in dialogue:** Discuss findings with peers to gain diverse perspectives and insights.
- **Use frameworks:** Apply frameworks like SWOT (Strengths, Weaknesses, Opportunities, and Threats) to analyze studies systematically.

Benefits of critical thinking

- Produces a literature review that is insightful and persuasive.
- Demonstrates a deep understanding of the research landscape.
- Lays a strong foundation for your research by highlighting its necessity and potential impact.

By incorporating critical thinking into the literature review, you not only enhance the quality and depth of your analysis but also position your research as a valuable contribution to the field.

3.4.2 Organization and structuring

A well-organized and logically structured literature review ensures clarity and coherence, enabling readers to understand the progression of ideas and the relevance of your research. Proper organization also highlights your critical analysis and synthesis of the existing body of knowledge.

Importance of organization and structuring

- **Enhances readability:** Makes it easier for readers to follow the flow of information.

- **Establishes context:** Clearly shows how your research fits within the existing literature.
- **Facilitates logical flow:** Presents ideas in a sequence that builds understanding and supports your argument.
- **Highlights key points:** Emphasizes critical findings and gaps in the literature.

Strategies for effective organization

- **Chronological order:** Arrange studies by publication date to show the evolution of knowledge in your field.
- **Thematic grouping:** Organize research by key themes or topics to highlight different aspects of the field.
- **Methodological approach:** Group studies based on the research methods used, such as qualitative, quantitative, or mixed methods.
- **Comparative analysis:** Present conflicting or complementary studies side-by-side for critical evaluation.

Components of a well-structured literature review

1. **Introduction**
    - Introduce the topic and its significance.
    - Outline the scope and criteria for selecting studies.
    - State the organization of the review (e.g., themes and chronology).
2. **Main Body**
    - Present studies in the chosen organizational structure.
    - Analyze and synthesize findings, emphasizing trends, gaps, and contradictions.
    - Use headings and subheadings for clarity.
3. **Conclusion**
    - Summarize key findings and their implications.
    - Highlight how your research addresses identified gaps.
    - Provide a transition to your research objectives or questions.

Example of organization

Thematic grouping example

- **Theme 1: Impact of social media on mental health**
  "Several studies have shown a correlation between social media use and increased anxiety levels (Author A, 2020; Author B, 2021)."
- **Theme 2: Positive effects of social media**
  "Conversely, some studies highlight the benefits of social media in fostering social connections (Author C, 2020; Author D, 2021)."

Chronological example

- 2000–2010: Early studies focused on the negative effects of excessive social media use.
- 2011–2020: Recent studies explored balanced usage and its nuanced impact.

By mastering the organization and structuring of your literature review, you ensure that your work is not only informative but also engaging and impactful, paving the way for a robust research contribution.

3.4.3 Summarization

Summarization is a critical skill in crafting a concise and focused literature review. It involves distilling large amounts of information into key points while maintaining the essence of the original content.

Key techniques

- **Identify core ideas:** Focus on the main arguments, findings, and conclusions of each source.
- **Use paraphrasing:** Rewrite information in your own words while preserving meaning.
- **Highlight contributions:** Emphasize how each work contributes to your research topic.
- **Avoid overloading:** Limit unnecessary details and tangential information.

Components of effective summarization

- **Relevance:** Include only information that aligns with your research focus.
- **Clarity:** Ensure the summary is straightforward and understandable.
- **Accuracy:** Represent the source material without misinterpretation.
- **Conciseness:** Keep it brief without losing the key message.

Steps to summarize

- Read the source thoroughly to understand its core content.
- Note the main points and their relevance to your study.
- Rewrite the points in a structured and simplified manner.
- Cross-check for accuracy and ensure proper citation.

Benefits

- Saves time for readers by eliminating redundant details.

- Enhances the flow and coherence of your literature review.
- Demonstrates your ability to synthesize and condense complex ideas.

By mastering summarization, you create a focused, engaging, and well-structured literature review that effectively communicates the foundation of your research.

3.4.4 Synthesis

Synthesis involves integrating information from multiple sources to identify patterns, relationships, and overarching themes. Unlike summarization, which focuses on condensing individual studies, synthesis combines ideas to create a cohesive narrative.

Key techniques

- **Group-related ideas:** Organize information from different studies into thematic clusters.
- **Compare and contrast:** Highlight similarities, differences, and contradictions across sources.
- **Establish connections:** Show how findings from various works build upon or challenge each other.
- **Identify trends:** Recognize recurring themes, methodologies, or gaps in the literature.

Components of effective synthesis

- **Relevance:** Focus on studies that directly relate to your research objectives.
- **Integration:** Seamlessly weave multiple sources into a unified discussion.
- **Critical analysis:** Evaluate the significance and reliability of the findings.
- **Original insight:** Provide your interpretation or perspective based on the synthesis.

Steps to synthesize

- **Organize sources:** Categorize studies based on common themes or methodologies.
- **Analyze relationships:** Identify how studies support, contradict, or complement each other.
- **Develop a narrative:** Create a logical flow that connects the grouped ideas.
- **Support claims:** Use evidence from the literature to substantiate your synthesis.

Benefits

- **Clarity:** Provides a clear understanding of the current state of research.
- **Insight:** Uncovers new perspectives or questions not addressed by individual studies.
- **Foundation:** Establish a strong basis for your research by contextualizing it within existing work.

By mastering synthesis, you transform a collection of isolated studies into a meaningful discussion, demonstrating your analytical abilities and deep understanding of the field.

3.4.5 Analytical skills

Analytical skills are essential for critically examining existing research and deriving meaningful insights. These skills enable writers to evaluate the quality, relevance, and contributions of individual studies and integrate them into a coherent narrative.

Importance of analytical skills

- **Evaluate credibility:** Assess the reliability of sources based on their methodology, authorship, and publication.
- **Identify gaps:** Recognize areas where further research is needed.
- **Discern patterns:** Highlight trends, inconsistencies, or evolving ideas within the field.
- **Assess relevance:** Ensure the literature selected directly supports the research objectives.

Key components of analytical skills

- **Critical thinking:** Question assumptions, methodologies, and conclusions of studies.
- **Comparative analysis:** Contrast findings to identify unique contributions or overlaps.
- **Logical reasoning:** Establish connections between studies and align them with your research focus.
- **Problem-solving:** Use insights from the literature to address research challenges or refine your objectives.

How to develop analytical skills for literature review

- **Engage deeply:** Read studies multiple times to fully understand their implications.

- **Ask questions:**
  1. What are the study's strengths and weaknesses?
  2. How does this work compare to others in the field?
  3. What assumptions underpin the findings?
- **Use tools:** Employ mind maps, charts, or tables to organize and analyze data from different sources.
- **Practice objectivity:** Avoid personal biases and focus on evidence-based evaluations.

Application in literature review

- **Critical evaluation:** Judge the validity of methodologies and the reliability of data.
- **Integration:** Synthesize findings to present a unified perspective.
- **Reflection:** Offer insights into how the literature informs or influences your research.

Benefits of analytical skills

- **Improved clarity:** Enhances the depth and quality of the literature review.
- **Better insights:** Identifies connections and gaps that might be overlooked.
- **Enhanced research impact:** Demonstrates your expertise and understanding of the field.

By honing analytical skills, you can craft a literature review that not only summarizes existing work but also provides a compelling foundation for your research.

3.4.6 Identifying gaps

Identifying gaps in the literature involves pinpointing areas that lack sufficient research or understanding within a specific field. This skill is critical for advancing knowledge and ensuring your research addresses unanswered questions or uncharted territory.

Importance of identifying gaps

- **Advances knowledge:** Contributes to the progression of the field by exploring overlooked areas.
- **Ensures relevance:** Aligns research with current needs and trends.
- **Avoids redundancy:** Prevents duplication of existing studies.
- **Highlights innovation:** Demonstrates the novelty and value of your research.

Steps to identify gaps in literature

- **Comprehensive review**
  1. Conduct a thorough analysis of existing studies, focusing on their objectives, methodologies, and findings.
- **Analyze trends**
  1. Identify recurring themes, patterns, or questions that remain unanswered.
- **Critique existing research**
  1. Evaluate the limitations and weaknesses of previous studies, such as small sample sizes or outdated methodologies.
- **Look for contradictions**
  1. Spot inconsistencies or conflicting results that need clarification.
- **Consider emerging topics**
  1. Explore areas influenced by new technologies, theories, or societal changes.

Examples of gaps

- **Methodological gaps:** Studies that fail to use advanced or diverse methodologies.
- **Contextual gaps:** Research conducted in limited geographic, cultural, or demographic contexts.
- **Theoretical gaps:** Lack of application or development of relevant theories.
- **Practical gaps:** Absence of real-world applicability or solutions to practical problems.

Techniques to highlight gaps

- **Compare studies**
  1. Contrast findings across multiple papers to identify areas of divergence or neglect.
- **Focus on future directions**
  1. Review the "future work" sections of papers for suggested areas of exploration.
- **Use systematic reviews**
  1. Employ systematic review techniques to ensure comprehensive coverage of the literature.
- **Consult experts**
  1. Seek insights from experienced researchers or practitioners in the field.

Benefits of identifying gaps

- **Strengthens research proposals:** Makes a compelling case for the significance of your study.
- **Promotes originality:** Ensures your work stands out in the academic or professional community.
- **Fosters collaboration:** Encourages partnerships by addressing shared knowledge deficits.

By effectively identifying gaps, you not only contribute to your field but also enhance the impact and relevance of your research. This process lays the foundation for innovation and meaningful contributions to knowledge.

3.5 SKILLS ARE NEEDED WHEN WRITING THE METHODS

The methods section is a critical part of any research paper, as it provides the details necessary for replicating the study and assessing its validity [63]. Writing this section effectively requires several specific skills to ensure accuracy, clarity, and completeness.

3.5.1 Clarity and precision

The methods section is the backbone of any research paper, detailing the processes, techniques, and tools used to achieve the study's objectives. Clarity and precision are critical to ensure that readers can replicate the study or understand its approach fully.

Importance of clarity and precision

- **Facilitates replicability:** Ensures that other researchers can accurately reproduce the study.
- **Enhances credibility:** Demonstrates a transparent and well-thought-out research process.
- **Aids comprehension:** Helps readers grasp complex procedures without ambiguity.

Key principles for achieving clarity and precision

- **Use simple and direct language**
    - Avoid overly technical jargon unless necessary.
    - Write in short, concise sentences to describe each step.
- **Provide specific details**
    - Specify the exact materials, tools, and instruments used (e.g., model numbers and software versions).

- Include precise measurements, conditions, and settings (e.g., temperature, pH levels, and time intervals).
- **Follow a logical order**
  - Present the methodology in the sequence it was performed.
  - Use subheadings to organize complex procedures into clear sections.
- **Be transparent**
  - Clearly describe all assumptions, controls, and limitations.
  - Specify how variables were defined, measured, and controlled.
- **Use visual aids where necessary**
  - Incorporate flowcharts, diagrams, or tables to summarize complex protocols.
  - Ensure visual aids are labelled clearly and referenced in the text.
- **Adhere to consistent terminology**
  - Use the same terms for tools, processes, and variables throughout the section.
  - Avoid synonyms that may confuse the reader.

Tips for ensuring clarity and precision

- **Be comprehensive but concise**
  - Include all necessary details but avoid overloading with redundant information.
- **Check for ambiguities**
  - Ensure there is no room for multiple interpretations of any step.
- **Anticipate reader questions**
  - Write as if explaining to someone unfamiliar with the study, addressing potential queries.
- **Peer review**
  - Have colleagues or mentors review the section to identify unclear or imprecise parts.

Example of clear and precise writing in methods

Less clear

- "We conducted experiments to measure the chemical properties of the solution."

More clear and precise

- "The pH levels of the solution were measured using a calibrated digital pH meter (Model XYZ-123) at room temperature (25°C). Measurements were taken at 15-minute intervals over a 2-hour period."

Benefits of clarity and precision

- **Replicability:** Other researchers can replicate the study, validating its results.
- **Transparency:** Builds trust in the research process and outcomes.
- **Professionalism:** Reflects the researcher's attention to detail and rigour.

Clarity and precision in the methods section not only strengthen the credibility of the research but also make it a valuable resource for future studies in the field.

3.5.2 Logical organization

The methods section must be systematically organized to present a coherent and accessible narrative of the research process. The logical organization ensures that readers can easily follow the steps, understand the rationale, and replicate the study if needed.

Importance of logical organization

- **Facilitates understanding:** Clear structure helps readers grasp the methodology without confusion.
- **Enhances replicability:** A well-organized section enables other researchers to duplicate the study.
- **Improves credibility:** Demonstrates a structured and systematic approach to research.

Key principles for logical organization

- **Chronological order**
  - Present steps in the order they were performed.
  - Use a timeline-like flow to prevent skipping or backtracking.
- **Group similar procedures**
  - Combine related methods into sections, such as "Data Collection," "Experimental Setup," and "Analysis."
  - Use subheadings to distinguish between different parts of the process.
- **Start with an overview**
  - Begin with a brief summary of the overall approach.
  - Include the rationale for selecting specific methods.
- **Use a standardized format**
  - Follow a consistent template, especially if guided by journal or institutional requirements.
  - Align with widely accepted practices in your discipline.

- **Highlight key steps**
  - Emphasize critical stages or unique aspects of the methodology.
  - Use bold text or numbering for clarity, if appropriate.
- **Provide context for each step**
  - Explain why each method or procedure was used.
  - Relate the methods to the research objectives.
- **Incorporate visual aids**
  - Use diagrams, flowcharts, or tables to present complex processes succinctly.
  - Place visual aids logically near the text that explains them.

Steps for logical organization

- **Introduction to methods**
  - Provide an overview of the methodology and its objectives.
- **Materials and tools**
  - List all equipment, materials, and software used, with specifications.
- **Step-by-step procedures**
  - Detail the steps in a logical, sequential order.
  - Include sufficient detail for replication.
- **Analysis and data processing**
  - Describe how data were analyzed, including statistical or computational tools.
- **Control measures and assumptions**
  - Specify how variables were controlled and assumptions handled.

Example of logical organization

Unorganized

- "The samples were prepared, and the data were analyzed. We then collected the materials, and statistical tools were applied."

Logically organized

- **Materials**
  - Sample collection involved using sterile containers (XYZ brand).
- **Sample preparation**
  - Samples were dried at 50°C for 24 hours in a controlled environment.
- **Data collection**
  - Measurements were recorded using an ABC spectrometer.
- **Analysis**
  - Statistical analysis was conducted using SPSS software (version X).

Benefits of logical organization

- **Reader engagement:** A structured flow makes the methods easier to read and understand.
- **Replicability:** Other researchers can follow the steps without ambiguity.
- **Professional Presentation:** Reflects a disciplined and methodical approach to research.

Logical organization in the methods section bridges the gap between complex research processes and clear communication, ensuring the research is both credible and impactful.

3.5.3 Consistency in terminology

Consistency in terminology is crucial in the methods section to maintain clarity, avoid confusion, and ensure the seamless communication of ideas. Using consistent terms throughout your paper fosters precision and enhances the reader's understanding.

Importance of consistency in terminology

- **Avoids misinterpretation:** Repeatedly using the same term ensures readers grasp its meaning without ambiguity.
- **Promotes clarity:** A consistent vocabulary eliminates the need for readers to deduce relationships between different terms.
- **Facilitates replication:** Consistent terminology makes it easier for other researchers to replicate your work accurately.

Best practices for maintaining consistency

- **Define key terms early**
  - Introduce and define all critical terms in the introduction or methods section.
  - Use these terms consistently throughout the document.
- **Avoid synonyms for key concepts**
  - Refrain from using multiple terms for the same concept, even if they are synonymous.
  - For instance, if you use "sample," do not interchange it with "specimen" or "test subject."
- **Standardize abbreviations and acronyms**
  - Define abbreviations the first time they are used (e.g., "Scanning Electron Microscopy (SEM)").
  - Use the abbreviation consistently afterwards.
- **Stick to discipline-specific terminology**
  - Use terms that are standard in your field and avoid introducing unnecessary jargon.

- **Align with journal or institutional guidelines**
  - Follow any specific terminological conventions provided by the journal or institution.
- **Use consistent units of measurement**
  - Maintain uniformity in the units used (e.g., always use "mg/L" rather than switching between "mg/L" and "ppm").
- **Review for terminology consistency**
  - During editing, check that the same terms are used consistently across all sections.

Example of inconsistency

Inconsistent terminology

- "The material was tested using SEM."
- "The specimen was analyzed using electron microscopy."

Consistent TERMINOLOGY

- "The sample was tested using Scanning Electron Microscopy (SEM)."

Tools for ensuring consistency

- **Style guides:** Refer to field-specific style guides or glossaries for standard terms.
- **Search function:** Use word processors to locate and replace inconsistent terms.
- **Collaborator feedback:** Ask co-authors or colleagues to review terminology consistency.

Benefits of consistency in terminology

- **Enhances readability:** Readers can focus on content without being distracted by varying terms.
- **Boosts credibility:** Consistency reflects professionalism and attention to detail.
- **Improves communication:** Clear terminology ensures that your message is understood across diverse audiences.

Consistency in terminology is a cornerstone of effective scientific writing, ensuring your research is accessible, reliable, and replicable.

3.5.4 Proper citation

Proper citation is a fundamental skill in academic writing, ensuring the acknowledgement of others' work while maintaining the integrity and credibility of your research. It involves accurately referencing sources to avoid plagiarism, provide evidence for claims, and guide readers to additional resources.

Importance of proper citation

- **Avoids plagiarism:** Proper citation gives credit to original authors, preventing the unethical practice of claiming others' work as your own.
- **Builds credibility:** Referencing authoritative sources strengthens the validity of your arguments.
- **Supports transparency:** Citations allow readers to verify sources and trace the origins of ideas or data.
- **Facilitates further research:** Accurate references guide readers to additional materials for deeper exploration.

Types of citations

- **In-text citations:** Brief references within the text (e.g., APA: "Smith, 2020").
- **Reference lists or bibliographies:** Detailed lists at the end of the document, including all cited works.
- **Footnotes/endnotes:** Additional information or citations are provided at the bottom of the page or end of the document.

Key principles of proper citation

- **Follow a citation Style**
  - Adhere to a recognized style (e.g., APA, MLA, Chicago, and IEEE) as required by the journal, institution, or discipline.
- **Include all necessary details**
  - Ensure citations contain all required elements, such as author names, publication year, title, and source details.
- **Cite all sources of information**
  - Acknowledge direct quotes, paraphrased ideas, data, and images from external sources.
- **Use consistent formatting**
  - Maintain uniformity in how citations are presented throughout the document.
- **Avoid overcitation and undercitation**
  - Strike a balance by citing only relevant and necessary sources.

Common citation styles

- **APA (American Psychological Association):** Used in social sciences.
  - Example: Smith, J. (2020). *Title of the Book*. Publisher.
- **MLA (Modern Language Association):** Used in humanities.
  - Example: Smith, John. *Title of the Book*. Publisher, 2020.

- **Chicago/turabian:** Used in history and business.
  - Example: Smith, John. *Title of the Book*. Publisher, 2020.
- **IEEE (Institute of Electrical and Electronics Engineers):** Used in technical fields.
  - Example: [1] J. Smith, *Title of the Book*, Publisher, 2020.

Best practices for proper citation

- **Use citation management tools**
  - Tools like Zotero, Mendeley, or EndNote can help organize references and generate citations.
- **Review citation guidelines**
  - Regularly consult the style manual to ensure adherence to rules.
- **Double-check sources**
  - Verify that all cited works appear in the reference list and vice versa.
- **Paraphrase thoughtfully**
  - Even when rephrasing, credit the original source.
- **Stay updated**
  - Citation styles are periodically updated; ensure you use the latest version.

Benefits of proper citation

- **Ethical research practice:** Demonstrates respect for intellectual property.
- **Enhanced collaboration:** Properly credited work fosters trust among researchers.
- **Professional presentation:** Reflects attention to detail and scholarly rigour.

Proper citation is not merely a formality but a cornerstone of scholarly communication, ensuring integrity, reliability, and respect within the academic and professional community.

3.5.5 Statistical and analytical understanding

Statistical and analytical understanding is an essential skill for researchers, enabling them to interpret data accurately, draw meaningful conclusions, and support their findings with evidence. This skill encompasses the ability to apply statistical tools, understand analytical methods, and evaluate results critically.

Importance in research

- **Validates findings:** Ensures the reliability and accuracy of conclusions through proper data analysis.

- **Supports decision-making:** Provides a basis for making informed decisions and hypotheses.
- **Identifies patterns:** Detects trends, correlations, and outliers in data sets.
- **Enhances credibility:** Demonstrates rigorous and methodical approaches to data interpretation.

Key aspects of statistical and analytical understanding

- **Understanding statistical tools**
  - Proficiency in software like SPSS, R, MATLAB, or Excel for data analysis.
  - Knowledge of basic statistical methods such as mean, median, standard deviation, and variance.
- **Hypothesis testing**
  - Applying tests, such as t-tests, ANOVA, or chi-square, to determine statistical significance.
- **Data interpretation**
  - Drawing conclusions from graphical representations like histograms, scatterplots, or box plots.
- **Regression analysis**
  - Using linear or logistic regression to explore relationships between variables.
- **Error analysis**
  - Understanding types of errors (Type I and Type II) and ensuring results are within acceptable error margins.

Analytical thinking in research

- **Problem-solving**
  - Breaking down complex problems into smaller, manageable components for analysis.
- **Critical evaluation**
  - Questioning the validity and reliability of data and methods used.
- **Pattern recognition**
  - Identifying trends and anomalies within data sets.
- **Decision-making**
  - Drawing logical and evidence-based conclusions.
- **Visualization**
  - Using charts, graphs, and infographics to represent data effectively.

Steps to develop statistical and analytical understanding

- **Learn basic statistics**
  - Master foundational concepts like probability, distributions, and hypothesis testing.

- **Understand analytical methods**
  - Study methods relevant to your field, such as qualitative, quantitative, or mixed methods.
- **Practice data analysis**
  - Work with real data sets to apply and test your statistical knowledge.
- **Stay updated**
  - Follow advancements in statistical techniques and software tools.
- **Seek feedback**
  - Collaborate with peers or mentors to refine your approach.

Applications in research

- **Survey analysis**
  - Interpreting responses to draw insights about populations or behaviours.
- **Experimental data**
  - Analyzing results to confirm or refute hypotheses.
- **Trend analysis**
  - Identifying long-term patterns in longitudinal studies.
- **Comparative studies**
  - Comparing groups to understand differences or similarities.
- **Predictive modeling**
  - Using statistical models to forecast outcomes.

Benefits of statistical and analytical understanding

- **Enhanced accuracy**
  - Reduces errors in interpretation and improves the reliability of conclusions.
- **Improved research quality**
  - Adds depth and rigour to studies, making them more impactful.
- **Better communication**
  - Helps convey findings effectively to stakeholders and peers.
- **Informed decisions**
  - Supports evidence-based policymaking and recommendations.

Statistical and analytical understanding is indispensable for any researcher, enabling them to transform raw data into meaningful insights while maintaining scientific integrity.

3.5.6 Brevity without omitting key details

Brevity in writing ensures that the content is concise, focused, and easy to understand, while still conveying all necessary information. It is a critical skill in research writing, particularly in sections like abstracts, summaries, and conclusions.

Importance of brevity

- **Enhances readability:** Keeps the audience engaged by avoiding unnecessary details.
- **Saves time:** Allows readers to grasp the main points quickly.
- **Improves clarity:** Prevents confusion caused by overly lengthy explanations.
- **Meets word limits:** Essential for journals, grants, or proposals with strict length requirements.

Strategies for achieving brevity

- **Prioritize information:** Focus on the most critical details that support your objectives.
- **Use precise language:** Replace verbose phrases with concise alternatives (e.g., "due to the fact that" → "because").
- **Avoid redundancy:** Eliminate repetitive statements or overlapping ideas.
- **Combine ideas:** Merge related points into a single, coherent sentence or paragraph.
- **Revise and edit:** Continuously refine your text to remove superfluous words or phrases.

Balancing brevity with completeness

- **Identify key details:** Ensure that all essential facts, findings, or arguments are included.
- **Avoid oversimplification:** Present complex ideas clearly without losing their depth or accuracy.
- **Use visuals:** Supplement text with charts, graphs, or tables to convey detailed information succinctly.

Examples of brevity

- **Verbose:** "The experiment was conducted over a period of seven days, during which observations were made and recorded meticulously."
- **Concise:** "The seven-day experiment included detailed observations."

Brevity enhances the impact of your writing, making it both efficient and effective while ensuring all vital information is retained.

3.6 SKILLS NEEDED WHEN WRITING THE RESULTS

The results section presents the findings of the research in a clear, concise, and unbiased manner [64]. It is essential to communicate data effectively without interpretation, leaving analysis and implications for the Discussion section. The following skills are crucial for writing an impactful results section:

3.6.1 Clarity in presentation

Clarity in presenting results ensures that findings are easily understandable, unambiguous, and aligned with the research objectives. It involves effectively organizing and communicating data to highlight its significance without misinterpretation.

Importance of clarity in results

- **Facilitates understanding:** Helps readers grasp the key outcomes of the study effortlessly.
- **Ensures transparency:** Clearly conveys how the data supports the research objectives and hypotheses.
- **Avoids misinterpretation:** Minimizes the risk of confusion by presenting findings systematically.
- **Strengthens credibility:** A well-structured presentation builds trust in the research's validity.

Key strategies for achieving clarity

- **Use clear language:** Avoid overly technical terms or jargon unless necessary.
  - Example: Replace "statistically significant" with "a meaningful difference" if explaining to a non-expert audience.
- **Logical structure:** Present results in a sequence that mirrors the research objectives or methodology.
  - Example: Group results by experiment, hypothesis, or variable for easy navigation.
- **Summarize key findings:** Highlight the most important results before delving into details.
- **Effective use of visuals:** Include tables, graphs, and charts to represent data concisely and visually.
  - Example: Use bar graphs to show comparisons or scatter plots for correlations.
- **Quantify findings:** Provide precise numerical values and avoid vague descriptors like "large" or "small."
  - Example: Instead of "a significant increase," write "a 25% increase."
- **Contextualize data:** Relate results to the research questions or hypotheses.
  - Example: "This 10% improvement supports Hypothesis 1, indicating enhanced efficiency."

Common mistakes to avoid

- **Overloading with data:** Presenting excessive details without prioritization can overwhelm readers.
- **Unclear visuals:** Poorly labeled graphs or tables hinder understanding.
- **Omitting negative results:** Failing to report unexpected or null findings reduces transparency.

Examples of clear results presentation

- **Unclear:** "The results showed differences between groups."
- **Clear:** "Group A outperformed Group B, with an average score of 85 compared to 75 ($p < 0.05$)."
- **Unclear visual:** A table with no labels or legend.
- **Clear visual:** A bar chart titled "Average Scores by Group," with axes labelled and a legend indicating group colours.

Clarity in presenting results not only enhances comprehension but also strengthens the overall impact of the research, making it accessible and valuable to its intended audience.

3.6.2 Logical organization

Logical organization ensures that research findings are presented in a structured and coherent manner, allowing readers to follow the flow of data and its implications seamlessly. It emphasizes systematic arrangement and clarity, aligning the results with the study's objectives and hypotheses.

Importance of logical organization in results

- **Enhances comprehension:** Readers can easily understand and interpret the findings.
- **Builds narrative flow:** Results logically follow the methodology and precede the discussion, maintaining consistency.
- **Facilitates comparisons:** Organized results enable clear identification of trends, patterns, and relationships.
- **Strengthens arguments:** A logical flow of results makes conclusions more persuasive and evidence-based.

Key elements of logical organization

- **Align with research objectives**
    - Present results in the same order as the objectives or hypotheses outlined in the introduction or methodology.
    - Example: If Objective 1 is about "impact on efficiency," present those results first.

- **Categorize findings**
  - Group related results together, such as experimental groups, variables, or time frames.
  - Example: Use subsections like "Group Comparisons" or "Time-Series Analysis."
- **Sequence for clarity**
  - Follow a natural progression: general trends first, followed by specific details or anomalies.
  - Example: Begin with a summary of key findings before diving into intricate data points.
- **Use headings and subheadings**
  - Break the section into logical parts with descriptive titles.
  - Example: "3.1 Key Trends," "3.2 Statistical Significance," and "3.3 Anomalous Observations."
- **Integrate visuals and text**
  - Place tables, charts, or graphs near the corresponding text to maintain context.
  - Example: Include a labelled graph immediately after discussing the related data.
- **Highlight key comparisons and contrasts**
  - Clearly outline similarities, differences, or changes observed in the data.
  - Example: "While Group A showed a 15% increase, Group B exhibited no significant change."

Common mistakes to avoid

- **Disjointed presentation:** Jumping between unrelated findings without transitions confuses readers.
- **Excessive details:** Overloading with raw data can obscure the main points.
- **Lack of prioritization:** Failing to emphasize the most critical findings diminishes the impact.
- **Inconsistent formatting:** Differing styles or layouts for presenting results reduce readability.

Common mistakes to avoid

- **Disjointed presentation:** Jumping between unrelated findings without transitions confuses readers.
- **Excessive details:** Overloading with raw data can obscure the main points.
- **Lack of prioritization:** Failing to emphasize the most critical findings diminishes the impact.
- **Inconsistent formatting:** Differing styles or layouts for presenting results reduce readability.

Logical organization ensures that the results section is not only informative but also intuitive to navigate. By structuring findings coherently, the research communicates its significance effectively and maintains the reader's engagement.

3.6.3 Use of visuals

Incorporating visuals in the results section enhances comprehension and allows complex data to be presented succinctly and effectively. Tables, graphs, charts, and other visual tools provide clarity, reinforce key findings, and help readers quickly grasp patterns and relationships.

Importance of visuals in the results section

- **Simplifies complex data:** Visuals condense detailed information into easily interpretable formats.
- **Enhances reader engagement:** Well-designed visuals capture attention and maintain interest.
- **Improves clarity:** Data trends, comparisons, and relationships are more apparent when presented visually.
- **Facilitates communication:** Universal formats like graphs transcend language barriers, making findings accessible to a broader audience.

Types of visuals and their applications

- **Tables**
    - Best for presenting precise numerical data or comparisons.
    - Example: Summarizing experimental results or demographic details.
- **Graphs**
    - Line Graphs: Ideal for showing trends over time.
    - Bar Graphs: Useful for comparing categories or groups.
    - Pie Charts: Effective for depicting proportions or percentages.
- **Charts**
    - Flowcharts: Explain processes or workflows.
    - Scatter Plots: Illustrate correlations between variables.
- **Images or Diagrams**
    - Include photographs, schematics, or conceptual diagrams to provide context or support findings.
- **Infographics**
    - Combine multiple visuals with brief explanations for a comprehensive overview.

Best practices for using visuals

- **Relevance**
    - Include visuals that directly support the findings and align with the research objectives.

    - Example: Use a graph to illustrate significant trends instead of overwhelming the section with raw data.
- **Clarity and simplicity**
    - Avoid overcrowding visuals with excessive information. Use clean designs with clear labels and legends.
    - Example: Ensure axes in graphs are labelled with units and descriptions.
- **Integration with text**
    - Place visuals close to the text that describes them to maintain context.
- **Consistency**
    - Maintain uniform formatting, such as font styles, colours, and axis labels, across all visuals.
    - Example: Use the same colour scheme for all bar graphs.
- **Captioning and referencing**
    - Provide descriptive captions for each visual, explaining its content and relevance.

Common mistakes to avoid

- **Overloading with visuals:** Too many visuals can overwhelm or confuse readers.
- **Inadequate labeling:** Missing or unclear labels reduce the interpretability of visuals.
- **Poor quality:** Blurry or poorly designed visuals undermine the credibility of the results.
- **Redundancy:** Avoid using visuals to repeat information already explained in detail.

Examples of visual usage

- **Before visuals**
    - "The experimental group showed a 20% improvement compared to the control group, which remained constant over three months."
- **With visuals**
    - A bar graph compares the experimental group's improvement with the control group over three months, making the trend immediately apparent.

Using visuals effectively in the results section not only enhances the presentation but also strengthens the overall impact of the research by making it accessible, engaging, and easy to understand.

3.6.4 Brevity with completeness

The ability to present results with brevity while ensuring completeness is a crucial skill for maintaining reader engagement and delivering clear, impactful findings. This balance allows for concise communication without omitting critical details that substantiate the research.

Key principles

- **Conciseness**
  - Present only relevant data that directly supports your research objectives.
  - Example: Instead of describing every step of data analysis, summarize the key outcomes.
- **Completeness**
  - Ensure all necessary details are included to provide a comprehensive understanding.
  - Example: Include sample sizes, statistical significance, and relevant metrics for context.
- **Avoiding redundancy**
  - Do not repeat information already mentioned in tables or visuals within the text.
  - Example: Instead of listing all data points, describe the trend or key insights.

Strategies for Brevity with Completeness

- **Prioritize key findings**
  - Focus on the most significant results related to your research questions.
  - Example: Highlight statistically significant differences rather than all tested variables.
- **Use summaries**
  - Summarize large datasets in tables or charts and provide a concise narrative in the text.
- **Integrate visual aids**
  - Leverage graphs and charts to convey complex data succinctly.
  - Example: Use a pie chart to show percentage distributions instead of lengthy textual descriptions.
- **Be specific and direct**
  - Avoid vague statements and provide precise figures and results.
  - Example: "The intervention improved scores by 25% ($p < 0.05$)," instead of "The intervention was effective."

Examples

- **Inefficient (wordy)**
  - "The analysis of the survey data indicated that a majority of respondents, which was about 65%, expressed satisfaction with the new system, while 20% were neutral, and the remaining 15% were dissatisfied."

- **Efficient (concise and complete)**
    - "Survey results showed 65% satisfaction, 20% neutrality, and 15% dissatisfaction with the new system."

Common pitfalls

- **Overloading with data**
    - Providing too many details can overwhelm readers. Instead, present highlights and refer to appendices for exhaustive data.
- **Omitting context**
    - Brevity should not compromise the clarity or completeness of the findings. Always include context where necessary.
- **Neglecting logical flow**
    - Ensure the results follow a logical progression to maintain coherence.

Benefits of brevity with completeness

- **Improves readability:** Concise text is easier for readers to follow and understand.
- **Maintains focus:** Keeps the discussion centred on critical findings without unnecessary distractions.
- **Enhances impact:** Highlights the significance of the results by eliminating superfluous information.

By adhering to these principles, researchers can present their results effectively, ensuring that the information is both accessible and authoritative.

3.6.5 Consistency in reporting

Consistency in reporting is vital for maintaining credibility, clarity, and readability in the results section of a paper. It ensures that data is presented uniformly and allows readers to easily follow and interpret the findings.

Key principles

- **Uniform terminology**
    - Use the same terms and phrases throughout the section to describe variables, groups, and outcomes.
    - Example: If using "control group" initially, avoid switching to "baseline group" later.
- **Standardized formats**
    - Present similar types of data in a consistent format, such as percentages, decimal places, or units of measurement.
    - Example: Report all p-values to two decimal places for uniformity.
- **Alignment with methodology**
    - Ensure that the results correspond to the methods described earlier in the paper.

    - Example: If you mentioned analyzing five variables, ensure results are reported for all five.
- **Logical sequence**
    - Follow a structured order, such as presenting results in the same sequence as the research questions or hypotheses.
    - Example: If hypothesis 1 is addressed first in the introduction, its results should appear first in this section.

Strategies for consistency in reporting

- **Establish reporting standards**
    - Use templates or guidelines for presenting results, especially in tables and figures.
    - Example: Apply consistent headers and labels across all visuals.
- **Review for repetition**
    - Cross-check that data is not redundantly repeated in multiple formats (e.g., in text and tables) without purpose.
    - Example: Avoid repeating numerical values verbatim if already shown in a table.
- **Align descriptions with data**
    - Match the narrative text with the figures, tables, or charts.
    - Example: If a graph shows "Group A's performance over time," the text should describe this trend using the same group name.
- **Check for numerical precision**
    - Use consistent decimal places, rounding methods, and units throughout.
    - Example: If reporting age as "25.3 years" in one part, avoid using "25 years" elsewhere.

Examples

- **Inconsistent reporting**
    - "The test scores increased by 15% in Group A, while Group B had a rise of 0.20."
- **Consistent reporting**
    - "The test scores increased by 15.0% in Group A, while Group B showed a rise of 20.0%."

Common pitfalls

- **Shifting terminology**
    - Using synonyms for key terms can confuse readers.
    - Example: Referring to "response time" as "reaction time" in different sections.
- **Mixed data presentation**
    - Switching between formats (e.g., percentages and raw numbers) without clarification.

- **Omission of context**
  - Not providing consistent background information to help readers interpret data.

Benefits of consistency in reporting

- **Enhances credibility:** Uniform reporting demonstrates professionalism and attention to detail.
- **Improves clarity:** A consistent approach helps readers understand and compare results more easily.
- **Facilitates peer review:** Reviewers can evaluate findings more effectively when data is presented consistently.

By ensuring consistency in reporting, researchers can strengthen the impact of their results and build trust with their audience.

3.6.6 Statistical accuracy

Statistical accuracy is critical in the results section to ensure that the data is reliable, valid, and correctly interpreted. It reflects the rigour of the research and supports the credibility of the conclusions drawn.

Key principles

- **Correct application of statistical tests**
  - Use appropriate statistical methods aligned with the research design and data type.
  - Example: Use a t-test for comparing means between two groups or ANOVA for more than two groups.
- **Proper representation of data**
  - Report data with precision, including measures of central tendency (mean and median) and variability (standard deviation and range).
  - Example: Instead of "the average was around 50," report "the mean was 50.3 (SD = 2.1)."
- **Transparency in reporting**
  - Include all relevant statistics, such as sample size, test statistic values, degrees of freedom, and p-values.
  - Example: "$t(28) = 2.34$, $p = 0.03$."
- **Avoid overgeneralization**
  - Do not make broad claims based on limited or insignificant statistical findings.
  - Example: Avoid stating "significant improvement" without clarifying the statistical basis.

Strategies for ensuring statistical accuracy

- **Double-check calculations**
  - Verify statistical computations using software or manual methods.
  - Example: Recheck confidence intervals to confirm they align with reported means and standard errors.
- **Follow reporting guidelines**
  - Adhere to standards like APA, MLA, or journal-specific formats for presenting statistics [65].
  - Example: In APA, report p-values as "$p < 0.05$" or exact values when appropriate.
- **Interpret results cautiously**
  - Provide clear distinctions between statistical significance and practical significance.
  - Example: A statistically significant result ($p < 0.05$) may not always imply real-world importance.
- **Avoid cherry-picking**
  - Present all relevant findings, including non-significant results, to provide a balanced view.
  - Example: If a secondary variable shows no significant effect, include this in the results section.

Examples

1. **Incorrect statistical reporting**
   - "The results showed an increase in performance, and it was significant."
2. **Accurate statistical reporting**
   - "The results showed a significant increase in performance ($t(34) = 3.56$, $p = 0.001$)."

Common pitfalls

1. **Misleading precision**
   - Over-reporting decimal places can imply false accuracy.
   - Example: Reporting $p = 0.0000001$ instead of $p < 0.001$.
2. **Incorrect test usage**
   - Applying a parametric test to non-parametric data or vice versa.
3. **Neglecting assumptions**
   - Ignoring assumptions like normality or homogeneity of variance required for certain tests.

Benefits of statistical accuracy

- **Enhances credibility:** Accurate statistics reflect the reliability and rigour of the research.

- **Supports valid conclusions:** Proper interpretation ensures that conclusions are based on solid evidence.
- **Facilitates peer review:** Accurate reporting allows reviewers to verify and evaluate findings effectively.

By maintaining statistical accuracy, researchers uphold the integrity of their work and contribute meaningfully to the scientific community.

3.6.7 Highlighting key findings

Highlighting key findings in the results section ensures that readers quickly grasp the most significant outcomes of your research [66]. This skill requires a balance between clarity, precision, and relevance to effectively communicate the essence of your study.

Key principles

- **Prioritize important results**
  - Focus on the findings that directly address the research objectives or hypotheses.
  - Example: Highlight significant correlations, trends, or patterns relevant to your study.
- **Use concise language**
  - Clearly and succinctly describe results without unnecessary elaboration.
  - Example: "The intervention group showed a 25% increase in performance compared to the control group (p = 0.02)."
- **Provide context**
  - Link findings to the research question to help readers understand their significance.
  - Example: "This result supports the hypothesis that X improves Y under condition Z."
- **Utilize visual aids**
  - Use tables, charts, or graphs to emphasize key results visually.
  - Example: Include a bar graph showing group comparisons or a scatter plot for correlations.
- **Avoid over-interpretation**
  - Present findings as they are without inferring implications that belong in the discussion section.
  - Example: State "Group A had higher scores than Group B" instead of "Group A's methods are superior."

Strategies for highlighting key findings

- **Organize results logically**

  - Present findings in a sequence that aligns with the research questions or hypotheses.
  - Example: Begin with primary results, followed by secondary or exploratory findings.
- **Use descriptive headings**
  - Break down results into sections with meaningful headings for clarity.
  - Example: "Effect of Treatment on Performance" or "Relationship Between Variables X and Y."
- **Summarize in bullet points**
  - For lengthy results, provide a summary of key findings in bullet points.
  - Example:
    1. The treatment group improved by 25% ($p = 0.02$).
    2. No significant difference was found in subgroup analysis ($p > 0.05$).
- **Emphasize visual cues**
  - Bold or italicize critical values, but use formatting sparingly to maintain professionalism.
  - Example: "The highest improvement was observed in Group A (mean = 45.3, **$p = 0.001$**)."

Common pitfalls

- **Overloading with data**
  - Avoid presenting every single result; focus on those most relevant to the research objectives.
- **Omitting key results**
  - Do not leave out important findings, even if they contradict your expectations.
- **Misleading presentation**
  - Ensure that highlighted findings accurately represent the data without exaggeration.

Benefits of highlighting key findings

- **Enhances reader comprehension:** Allows readers to quickly identify the study's main contributions.
- **Facilitates decision-making:** Helps reviewers, stakeholders, or practitioners understand the relevance of your work.
- **Improves overall impact:** This makes the research more memorable and impactful by emphasizing its core results.

By mastering the art of highlighting key findings, you effectively showcase the value and significance of your research, ensuring its contributions are clearly understood and appreciated.

3.7 SKILLS ARE NEEDED WHEN WRITING THE DISCUSSION

The discussion section is where you interpret and contextualize your results, linking them to your research objectives and the existing body of knowledge [67]. It demonstrates the significance of your findings, explores implications, and acknowledges limitations. Writing a compelling discussion requires the following key skills:

3.7.1 Critical thinking

Critical thinking is fundamental in crafting a compelling and insightful discussion section. It involves analyzing results, interpreting their implications, and connecting findings to existing knowledge, all while maintaining objectivity and depth.

Key aspects of critical thinking in the discussion

- **Analyzing results**
    - Examine your findings in detail to uncover patterns, relationships, or anomalies.
    - Example: "The unexpected decline in efficiency at higher temperatures suggests potential interference from unaccounted variables."
- **Interpreting implications**
    - Assess what the results mean for the research question and broader field.
    - Example: "These results imply that the proposed method is more effective than traditional techniques under low-light conditions."
- **Comparing with existing literature**
    - Critically evaluate how your findings align with, contradict, or extend previous studies.
    - Example: "Unlike Smith et al. (2020), our results show no significant effect of variable X on outcome Y, suggesting the role of context-specific factors."
- **Acknowledging limitations**
    - Reflect on the study's constraints and how they might affect interpretations.
    - Example: "The small sample size may limit the generalizability of these findings."
- **Proposing future directions**
    - Suggest avenues for further research based on identified gaps or unresolved questions.
    - Example: "Future studies could explore the long-term effects of intervention Z in diverse populations."

Strategies to apply critical thinking

- **Question assumptions**
  - Scrutinize underlying premises in your analysis and interpretations.
  - Example: "Does the observed trend hold true across all demographic groups, or is it specific to this sample?"
- **Draw evidence-based conclusions**
  - Base your interpretations on solid evidence rather than speculation.
  - Example: "The consistent increase in yield supports the hypothesis that nutrient supplementation improves growth rates."
- **Explore alternative explanations**
  - Consider other factors that might explain your results.
  - Example: "The observed variability might also result from measurement errors rather than inherent differences."
- **Integrate broader context**
  - Relate findings to real-world applications or theoretical frameworks.
  - Example: "These insights could improve policy decisions in renewable energy management."
- **Be objective and balanced**
  - Avoid overstating results or ignoring contradictory evidence.
  - Example: "While the findings are promising, they must be interpreted cautiously given the study's constraints."

Common pitfalls to avoid

- **Overgeneralization**
  - Drawing broad conclusions from limited data.
  - Example: "Claiming a universal solution based on results from a single region is misleading."
- **Ignoring contradictions**
  - Failing to address discrepancies with previous studies.
  - Example: "Neglecting conflicting evidence reduces the credibility of the discussion."
- **Speculation without evidence**
  - Making claims unsupported by data.
  - Example: "Assuming causation where only correlation is observed."
- **One-sided analysis**
  - Highlighting only favourable outcomes while ignoring limitations or weaknesses.
  - Example: "Focusing solely on successful trials without addressing failed attempts."

Benefits of critical thinking in the discussion

- **Enhances depth:** Provides a nuanced and comprehensive understanding of results.

- **Builds credibility:** Demonstrates analytical rigor and transparency.
- **Encourages engagement:** Invites readers to think critically about the study's implications.
- **Informs Future research:** Lays a strong foundation for subsequent investigations.

By employing critical thinking, the discussion becomes a platform to showcase the intellectual merit of the research, positioning it as a meaningful contribution to the field.

3.7.2 Contextualizing results

Contextualizing results involves situating your findings within the broader landscape of existing knowledge, real-world applications, or theoretical frameworks. This skill ensures that your research is meaningful and relatable and contributes to the ongoing discourse in your field.

Key aspects of contextualizing results

- **Linking to research objectives**
  - Reiterate how your findings address the research questions or objectives.
  - Example: "The results confirm our hypothesis that increased nutrient availability enhances plant growth, aligning with the study's primary goal."
- **Connecting to existing literature**
  - Compare and contrast your results with previous studies to highlight their significance.
  - Example: "These findings corroborate the work of Johnson et al. (2019), who reported similar trends in urban environments."
- **Relating to real-world applications**
  - Discuss how the findings can be applied in practical scenarios or influence policy.
  - Example: "This method could improve water management practices in arid regions by optimizing resource allocation."
- **Addressing broader implications**
  - Explore how the results contribute to theoretical frameworks or societal challenges.
  - Example: "These insights extend our understanding of behavioural economics, particularly in decision-making under uncertainty."
- **Highlighting novel contributions**
  - Emphasize how your study fills gaps or introduces new perspectives.
  - Example: "This study is the first to demonstrate the role of microhabitats in shaping species interactions."

Strategies for contextualizing results

- **Engage with key literature**
  - Reference foundational studies and recent advancements to position your work.
  - Example: "Building on Smith's (2015) model, our findings introduce a novel variable that explains additional variance."
- **Use theoretical frameworks**
  - Situate findings within established theories or models.
  - Example: "These results align with the diffusion of innovation theory, suggesting early adopters drive the observed trends."
- **Consider practical impacts**
  - Discuss how your findings could solve problems or create opportunities.
  - Example: "Implementing this technology in healthcare could reduce diagnostic errors by 20%."
- **Address contradictions**
  - Contextualize discrepancies by exploring potential reasons for divergence.
  - Example: "Unlike prior studies, our results suggest a weaker correlation, possibly due to differences in sample demographics."
- **Bridge knowledge gaps**
  - Explicitly state how your findings contribute to areas of uncertainty or limited research.
  - Example: "This study bridges the gap between laboratory experiments and field applications, providing actionable insights."

Common pitfalls to avoid

- **Lack of relevance**
  - Failing to tie findings to the research question or broader context.
  - Example: "Discussing results in isolation diminishes their impact and significance."
- **Overstating implications**
  - Making claims that extend beyond the scope of the study.
  - Example: "Assuming global applicability based on a single case study undermines credibility."
- **Ignoring contradictory evidence**
  - Failing to address how your findings differ from or challenge existing research.
  - Example: "Overlooking opposing studies leaves gaps in the discussion."
- **Vague generalizations**
  - Providing overly broad statements without specifics.
  - Example: "Claiming 'significant contributions' without detailing their nature or scope."

Benefits of contextualizing results

- **Enhances understanding:** Provides a comprehensive view of the research's relevance.
- **Builds credibility:** Demonstrates thorough knowledge of the field.
- **Encourages engagement:** Helps readers connect with the study's implications.
- **Drives future research:** Sparks ideas for further exploration or application.

By effectively contextualizing results, the discussion becomes a meaningful bridge between data and its broader significance, showcasing the study's value to both academic and practical domains.

3.7.3 Speculating reasonably

Speculating reasonably in the discussion section involves interpreting findings and proposing plausible explanations or implications without overstepping the evidence. This skill requires a balance between creativity and caution, ensuring that speculation is informed, logical, and grounded in data.

Key aspects of speculating reasonably

- **Grounded in evidence**
  - Base speculation on the results of your study and existing literature.
  - Example: "The observed increase in yield may be attributed to enhanced nutrient absorption, as suggested by prior studies on similar soil types."
- **Acknowledging uncertainty**
  - Clearly indicate when a statement is a hypothesis or assumption.
  - Example: "It is possible that unmeasured variables, such as local microclimate conditions, contributed to the unexpected findings."
- **Maintaining relevance**
  - Focus speculation on aspects directly related to your research questions or objectives.
  - Example: "These findings might inform future studies on sustainable urban planning, particularly in developing regions."
- **Avoiding overgeneralization**
  - Limit speculation to the scope of your data and study context.
  - Example: "While the results are promising, further research is needed to determine their applicability to larger populations."
- **Proposing future research**
  - Use speculation to suggest areas for further investigation.
  - Example: "Further experiments could explore the role of light intensity in modulating this effect, providing deeper insights."

Strategies for speculating reasonably

- **Stay aligned with findings**
  - Ensure that all speculative statements can be logically traced back to the data.
  - Example: "Given the correlation observed, it is plausible that economic incentives play a key role in behaviour modification."
- **Cite supporting literature**
  - Use existing studies to justify or support speculative ideas.
  - Example: "This hypothesis aligns with findings by Lee et al. (2020), who noted similar patterns in comparable ecosystems."
- **Incorporate multiple perspectives**
  - Consider alternative explanations or viewpoints to strengthen credibility.
  - Example: "Another possible explanation is that seasonal variations influenced the results, though this requires confirmation."
- **Use conditional language**
  - Employ terms like "might," "could," or "potentially" to indicate speculative nature.
  - Example: "These findings might suggest an underlying genetic mechanism, though further molecular studies are required."
- **Propose practical applications thoughtfully**
  - Discuss real-world implications cautiously, ensuring they are realistic.
  - Example: "If confirmed in larger trials, this approach could potentially reduce energy consumption in industrial processes."

Common pitfalls to avoid

- **Overstating certainty**
  - Presenting speculative ideas as definitive conclusions undermines credibility.
  - Example: "Claiming causation without experimental control misrepresents the data."
- **Ignoring study limitations**
  - Failing to acknowledge the constraints of your study when speculating.
  - Example: "Speculating beyond the geographic or demographic scope of the research leads to overreach."
- **Vague or unsupported speculation**
  - Making broad statements without evidence or logical reasoning.
  - Example: "Suggesting outcomes without tying them to data or literature weakens the argument."
- **Excessive speculation**
  - Overloading the discussion with hypothetical scenarios distracts from core findings.

- Example: "Providing too many 'what ifs' can dilute the focus and impact of the study."

Benefits of reasonable speculation

- **Encourages innovation:** Opens pathways for new ideas and hypotheses.
- **Engages the audience:** Invites readers to think critically about the implications.
- **Drives future research:** Suggests meaningful directions for continued exploration.
- **Enhances impact:** Demonstrates thoughtful interpretation of results.

By speculating reasonably, you enhance the discussion's depth and relevance while maintaining scientific rigour, making your research a valuable contribution to the field.

3.7.4 Identifying future research directions

Identifying future research directions is a vital skill that demonstrates the broader impact of your study and guides subsequent work in the field [68]. It involves pinpointing unanswered questions, proposing extensions of the current research, and suggesting new methodologies or areas of application.

Key aspects of identifying future research directions

- **Highlighting research gaps**
  - Identify areas that your study was unable to address.
  - Example: "While this study focused on short-term outcomes, future research could investigate the long-term effects of this intervention."
- **Proposing new questions**
  - Formulate research questions that arise from your findings.
  - Example: "What role do individual behavioural differences play in the observed trends?"
- **Suggesting methodological improvements**
  - Recommend enhanced techniques or alternative approaches.
  - Example: "Future studies might benefit from incorporating real-time monitoring systems to increase data accuracy."
- **Exploring broader applications**
  - Suggest ways to apply the findings in different contexts or fields.
  - Example: "The model could be adapted for use in urban planning to optimize energy consumption."
- **Addressing limitations**
  - Turn the limitations of your study into opportunities for future exploration.
  - Example: "Due to resource constraints, this study was limited to a small sample size; larger-scale studies could provide more generalizable insights."

Strategies for identifying future research directions

- **Connect to your findings**
  - Ensure suggestions for future research logically follow from your results.
  - Example: "Given the strong correlation between X and Y, further research could explore causative mechanisms."
- **Incorporate emerging trends**
  - Align future research suggestions with current advancements in the field.
  - Example: "With the advent of AI, future studies could utilize machine learning to refine predictive models."
- **Draw on interdisciplinary insights**
  - Suggest exploring intersections with other disciplines.
  - Example: "Combining insights from neuroscience and computer science could enhance understanding in this area."
- **Be specific and actionable**
  - Provide clear, focused suggestions rather than vague ideas.
  - Example: "Future research should investigate the role of temperature variations in the observed chemical reaction rates."
- **Prioritize impactful areas**
  - Focus on research directions that address pressing issues or have significant potential benefits.
  - Example: "Investigating this mechanism further could contribute to sustainable energy solutions."

Common pitfalls to avoid

- **overgeneralization**
  - Suggesting overly broad research directions that lack focus.
  - Example: "Simply stating that 'more research is needed' without specifics fails to provide value."
- **Ignoring study limitations**
  - Overlooking how limitations inform future research.
  - Example: "Not addressing a limitation such as sample diversity misses an opportunity for constructive suggestions."
- **Being overly ambitious**
  - Proposing research directions that are impractical or beyond the current scope of the field.
  - Example: "Suggesting global-scale experiments without considering feasibility reduces credibility."
- **Repetition of existing ideas**
  - Reiterating previously explored directions without adding new insights.
  - Example: "Proposing research that merely replicates existing studies adds little value."

Benefits of identifying future research directions

- **Drives innovation:** Encourages the exploration of novel ideas and methodologies.
- **Promotes collaboration:** Inspires interdisciplinary or collaborative projects.
- **Demonstrates thought leadership:** Highlights your ability to contribute meaningfully to the field.
- **Guides the research community:** Provides a roadmap for addressing critical issues.

By thoughtfully identifying future research directions, you enhance the relevance and impact of your work, ensuring it serves as a foundation for continued progress in the field.

3.7.5 Integrating evidence

Integrating evidence in the discussion section is critical for connecting your findings to existing knowledge, supporting your interpretations, and strengthening your arguments. This skill involves effectively blending data, theories, and prior research to provide a comprehensive and credible analysis.

Key elements of integrating evidence

- **Connecting to findings**
  - Relate your evidence directly to the results of your study.
  - Example: "The observed increase in efficiency aligns with previous studies that identified X as a critical factor."
- **Referencing relevant literature**
  - Cite studies that support or contrast your findings to contextualize your research.
  - Example: "Unlike Smith et al. (2020), who reported no significant effect, our results indicate a strong correlation between Y and Z."
- **Using data effectively**
  - Incorporate quantitative and qualitative data to substantiate claims.
- **Synthesizing information**
  - Combine multiple sources to highlight patterns, contrasts, or trends.
  - Example: "The consistency across studies suggests a robust relationship, although differences in methodology might explain the observed variations."
- **Addressing contradictions**
  - Acknowledge conflicting evidence and provide possible explanations.

- Example: "While prior research suggests X, the current study's findings may differ due to Y factors."

Strategies for integrating evidence

- **Prioritize relevance**
  - Select evidence that directly supports the points being discussed.
  - Example: "Focusing on studies conducted under similar conditions ensures meaningful comparisons."
- **Use clear and concise language**
  - Avoid overloading sentences with excessive references or data.
  - Example: "The study confirms earlier findings (Author, Year) while introducing new perspectives on Y."
- **Maintain logical flow**
  - Present evidence in a sequence that builds your argument effectively.
  - Example: "Start with broad agreements, then move to specific divergences to guide the reader through your analysis."
- **Incorporate visual aids**
  - Use tables, charts, or figures to present evidence succinctly.
- **Balance supporting and opposing evidence**
  - Present a fair view by discussing evidence that both supports and challenges your findings.
  - Example: "While most studies agree on the positive effects of X, some evidence suggests potential limitations under Y conditions."

Common pitfalls to avoid

- **Cherry-picking evidence**
  - Only presenting evidence that supports your conclusions while ignoring contrary data.
  - Example: "This undermines the credibility of your discussion and leaves gaps in your argument."
- **Overloading with references**
  - Including too many citations without clear connections to your findings.
  - Example: "A cluttered discussion makes it difficult for readers to discern the key takeaways."
- **Failing to interpret evidence**
  - Merely listing studies without analyzing their implications.
  - Example: "Simply stating, 'Studies A, B, and C report similar findings,' does not provide insight."
- **Misrepresenting data**
  - Drawing conclusions not supported by the evidence.
  - Example: "Exaggerating the significance of minor correlations can mislead readers."

Benefits of effective evidence integration

- **Enhances credibility:** Demonstrates a thorough understanding of the field and supports your interpretations.
- **Strengthens arguments:** Provides a solid foundation for your conclusions.
- **Encourages critical thinking:** Highlights your ability to analyze and synthesize complex information.
- **Facilitates knowledge building:** Connects your study to the broader body of research, promoting scientific progress.

By mastering the integration of evidence, you ensure your discussion is both persuasive and grounded, making your research more impactful and reliable.

3.8 SKILLS ARE NEEDED WHEN WRITING THE CONCLUSIONS

The conclusions section is the final opportunity to summarize your research, emphasize its significance, and suggest future directions [69]. Writing a compelling and concise conclusion requires the following key skills:

3.8.1 Summarizing key findings

Summarizing key findings in the conclusion section is an essential skill that encapsulates the essence of your research. It ensures that readers leave with a clear understanding of the study's outcomes and their significance.

Key aspects of summarizing key findings

- **Focus on core results**
  - Highlight the most important findings that directly address your research objectives.
  - Example: "The study confirmed a 25% improvement in efficiency using method X, supporting its applicability in Y scenarios."
- **Avoid repetition**
  - Summarize findings without duplicating the detailed explanations provided in the results or discussion sections.
  - Example: "Rather than reiterating the entire dataset, emphasize the trends and key observations."
- **Maintain clarity**
  - Use concise and straightforward language to present findings.
  - Example: "The analysis revealed a strong correlation between A and B, consistent across all conditions."

- **Connect to objectives**
  - Relate findings back to the research objectives stated in the introduction.
  - Example: "The results address the primary goal of understanding the impact of Z on Y, confirming the initial hypothesis."
- **Highlight novel contributions**
  - Emphasize how your findings contribute to the field or fill existing gaps.
  - Example: "This study is the first to demonstrate the effect of X under Y conditions, providing new insights into Z."

Strategies for summarizing key findings

- **Prioritize relevance**
  - Include only the findings that directly relate to the research questions or objectives.
  - Example: "Focus on outcomes that have significant implications for the field."
- **Use thematic grouping**
  - Organize findings under themes or categories to enhance clarity.
  - Example: "Group findings related to methodology, outcomes, and implications separately."
- **Be specific and quantitative**
  - Provide specific results or data points where possible.
  - Example: "The study found a 30% reduction in energy consumption, validating the efficiency of the proposed method."
- **Integrate with broader implications**
  - Link findings to their potential applications or theoretical significance.
  - Example: "These results suggest practical applications in X and contribute to theoretical models of Y."
- **Avoid overgeneralization**
  - Ensure conclusions are directly supported by your data and analysis.
  - Example: "Avoid broad claims not substantiated by the findings."

Common pitfalls to avoid

- **Overloading with details**
  - Summarizing is about distilling, not repeating every detail.
  - Example: "Focus on the 'what' and 'why,' not the 'how' from earlier sections."
- **Neglecting negative findings**
  - Address findings that did not align with expectations as they add value to the research narrative.
  - Example: "The unexpected result in condition B highlights the need for further exploration."

- **Failing to connect to objectives**
  - A disjointed summary weakens the impact of your findings.
  - Example: "Tie conclusions explicitly to the study's aims for coherence."
- **Being vague**
  - Ambiguous statements reduce the clarity and utility of the conclusions.
  - Example: "Avoid saying 'some improvement'—quantify or qualify the improvement."

Benefits of effective summarization

- **Enhances reader comprehension:** Provides a clear takeaway of the study's achievements.
- **Reinforces research impact:** Underscores the significance of the findings in addressing the research question.
- **Guides future work:** Suggests avenues for further research based on the summarized outcomes.
- **Supports practical application:** Highlights actionable insights derived from the study.

By mastering the skill of summarizing key findings, researchers can effectively convey the essence of their work, leaving a lasting impression on readers and stakeholders.

3.8.2 Offering practical implications

Offering practical implications in the conclusion section bridges the gap between theoretical research and real-world applications. It ensures that the findings are not only academically valuable but also relevant and actionable for practitioners, policymakers, or other stakeholders.

Key aspects of offering practical implications

- **Highlight relevance to real-world problems**
  - Connect findings to current challenges or needs in the field.
  - Example: "This method can reduce production costs in manufacturing industries by 20%, making it a viable solution for cost optimization."
- **Focus on specific applications**
  - Clearly define how the findings can be implemented in practice.
  - Example: "The developed algorithm can be integrated into smart grid systems to enhance energy distribution efficiency."
- **Consider the stakeholder perspective**
  - Address how different stakeholders (e.g., industries, communities, or policymakers) can benefit.
  - Example: "Policy makers can use these insights to draft more effective environmental regulations."

- **Balance innovation and feasibility**
  - Propose ideas that are both novel and achievable within existing constraints.
  - Example: "Adopting this strategy requires minimal infrastructure changes, making it a cost-effective solution."
- **Address limitations and context**
  - Mention conditions under which the practical implications are most applicable.
  - Example: "These results are most relevant to urban areas with high traffic congestion."

Strategies for offering practical implications

- **Identify the core takeaways**
  - Pinpoint findings that have the most significant practical relevance.
  - Example: "Focus on outcomes that directly address operational challenges in the field."
- **Use action-oriented language**
  - Frame implications as actionable steps or recommendations.
  - Example: "Organizations should consider adopting X technology to enhance operational efficiency."
- **Provide examples**
  - Use case studies or scenarios to illustrate how the findings can be applied.
  - Example: "For instance, implementing this system in city Y reduced energy consumption by 15%."
- **Incorporate stakeholder feedback**
  - Reflect on how stakeholders perceive the utility of your findings.
  - Example: "Preliminary feedback from practitioners indicates strong potential for scaling this approach."
- **Link to broader goals**
  - Connect implications to societal, economic, or environmental objectives.
  - Example: "This approach aligns with global sustainability goals by reducing carbon emissions."

Common pitfalls to avoid

- **Overgeneralizing applications**
  - Avoid suggesting universal applicability without considering specific contexts.
  - Example: "State where and how the findings are applicable rather than implying they work everywhere."
- **Neglecting feasibility**
  - Unrealistic recommendations reduce credibility.

    - Example: "Consider resource and technical constraints when proposing solutions."
- **Failing to engage stakeholders**
    - Overlooking the needs and priorities of end-users diminishes practical relevance.
    - Example: "Tailor implications to address stakeholder-specific challenges."
- **Omitting limitations**
    - Practical implications should acknowledge any limitations of the study.
    - Example: "State that the method works best in conditions A and B but may require adjustments for C."

Benefits of offering practical implications

- **Enhances impact:** Demonstrates the real-world value of the research.
- **Attracts broader interest:** Engages a wider audience beyond academia.
- **Guides action:** Provides clear pathways for implementing findings.
- **Encourages collaboration:** Opens opportunities for partnerships with industries, governments, and other organizations.

By offering well-thought-out practical implications, researchers ensure their work extends beyond theoretical insights to drive meaningful change and innovation.

3.8.3 Proposing future research

Proposing future research in the conclusion provides a pathway for advancing knowledge in the field. It highlights areas that remain unexplored, unresolved questions, or new avenues arising from the study. This section demonstrates the broader relevance of your work and encourages continuity in the academic conversation.

Key Aspects of Proposing Future Research

- **Identifying knowledge gaps**
    - Point out limitations or unanswered questions from the current study.
    - Example: "Further investigation is needed to explore the long-term effects of this intervention on diverse populations."
- **Building on current findings**
    - Suggest extensions or applications of your results to other contexts or fields.
    - Example: "Future studies could test this framework in different industries to evaluate its adaptability."

- **Suggesting methodological improvements**
  - Recommend enhancements in research design or tools for future studies.
  - Example: "Using advanced imaging techniques could provide more detailed insights into this phenomenon."
- **Highlighting emerging trends**
  - Encourage research that aligns with current or future trends in the field.
  - Example: "As artificial intelligence continues to evolve, integrating it into this approach could yield transformative results."
- **Encouraging multidisciplinary approaches**
  - Advocate for collaborations across disciplines to tackle complex problems.
  - Example: "Combining insights from materials science and mechanical engineering could lead to innovative solutions."

Strategies for proposing future research

- **Be specific and focused**
  - Avoid vague suggestions; clearly define the scope of future studies.
  - Example: "Future research should examine the effect of variable X on outcome Y under condition Z."
- **Address practical relevance**
  - Emphasize how the proposed research could solve real-world problems.
  - Example: "Exploring this model's application in renewable energy could address critical sustainability challenges."
- **Incorporate feasibility**
  - Ensure the proposed ideas are achievable with available resources and technology.
  - Example: "A longitudinal study with moderate sample sizes could effectively address this gap."
- **Leverage limitations as opportunities**
  - Turn study constraints into a basis for future exploration.
  - Example: "The small sample size in this study suggests a need for larger-scale trials to confirm these findings."
- **Align with broader goals**
  - Connect proposed research to global or field-specific priorities.
  - Example: "This direction aligns with the United Nations Sustainable Development Goals."

Common pitfalls to avoid

- **Overgeneralization**
  - Avoid proposing overly broad or undefined research areas.

    - Example: "Suggest specific variables, populations, or methods to investigate."
- **Ignoring Feasibility**
    - Proposals should be realistic given current knowledge and resources.
    - Example: "Focus on attainable objectives within practical constraints."
- **Neglecting the field's needs**
    - Avoid proposing research that lacks relevance or impact.
    - Example: "Prioritize areas with significant theoretical or practical importance."
- **Overlooking ethical considerations**
    - Ensure proposed research adheres to ethical standards.
    - Example: "Consider participant privacy and data security in future studies."

Benefits of proposing future research

- **Advances the field:** Encourages new studies that expand the scope of knowledge.
- **Engages the academic community:** Inspires collaboration and dialogue among researchers.
- **Demonstrates critical thinking:** Reflects a thorough understanding of the study's implications.
- **Increases relevance:** Shows the study's contribution to ongoing scientific and practical challenges.

By effectively proposing future research, authors not only reinforce the importance of their findings but also contribute to the dynamic evolution of their field.

3.8.4 Avoiding repetition

Avoiding repetition in the conclusion ensures that the final section remains impactful and does not simply restate what has already been covered. Instead, it should synthesize the study's findings and emphasize their significance without duplicating earlier content.

Key aspects of avoiding repetition

- **Paraphrase key points**
    - Reframe the core findings in fresh language to maintain reader engagement.
    - Example: Instead of repeating, "This study highlights the effectiveness of X," rephrase as "The findings underscore the practical benefits of X in achieving Y."

- **Focus on implications**
  - Shift from summarizing results to discussing their broader significance.
  - Example: "These results suggest practical applications in the fields of healthcare and technology."
- **Introduce a forward-looking perspective**
  - Highlight future directions or unanswered questions rather than reiterating past details.
  - Example: "Building on these findings, future research can explore…"
- **Summarize, don't repeat**
  - Provide a concise synthesis of the study without rehashing entire sections.
  - Example: "This research confirms the hypothesis and reveals new insights into…"

Strategies to avoid repetition

- **Use synonyms and alternative phrasing**
  - Replace repetitive terms with synonyms or restructured sentences.
  - Example: Instead of repeatedly saying "important findings," use "significant outcomes," or "key discoveries."
- **Prioritize new information**
  - Reserve the conclusion for implications, applications, or forward-looking statements.
  - Example: Avoid re-explaining the methodology or re-listing results.
- **Integrate multiple points**
  - Combine related findings into a single, cohesive statement.
  - Example: "Collectively, these results highlight both the challenges and opportunities in…"
- **Tailor content to the audience**
  - Focus on what is most relevant and valuable to the reader at this stage.
  - Example: Use the conclusion to address the "so what?" of your research.

Common pitfalls to avoid

- **Restating the abstract**
  - Avoid summarizing the entire paper again. The abstract already serves this purpose.
- **Repeating entire sections**
  - Refrain from copying text verbatim from the results or discussion sections.
- **Overemphasizing one point**
  - Maintain a balanced overview without dwelling excessively on a single finding.

- **Using redundant phrases**
  - Eliminate phrases like "as mentioned earlier" or "as discussed previously."

Benefits of avoiding repetition

- **Enhances readability:** Keeps the conclusion concise and engaging.
- **Improves professionalism:** Demonstrates effective communication skills.
- **Maximizes impact:** Ensures the conclusion leaves a lasting impression on the reader.
- **Maintains reader interest:** Avoids redundancy, keeping the reader focused on the significance of the research.

By crafting a conclusion that synthesizes and expands upon the study's contributions, authors can deliver a compelling and memorable final message.

3.8.5 Ending on a strong note

Ending on a strong note ensures the conclusion leaves a lasting impression, reinforces the significance of the research, and provides closure to the reader. A powerful ending highlights the study's contributions, underscores its broader implications, and inspires further thought or action.

Key strategies for ending on a strong note

- **Reaffirm the study's impact**
  - Emphasize the significance of your findings and their relevance to the field.
  - Example: "This study provides a foundational understanding of X, offering valuable insights for both theoretical advancement and practical applications."
- **Connect back to the introduction**
  - Tie the conclusion to the research question or problem statement introduced at the beginning.
  - Example: "By addressing the challenges outlined initially, this research paves the way for…"
- **Offer a broader perspective**
  - Situate your findings within a larger context or field of study.
  - Example: "These findings contribute to the growing body of knowledge on Y, highlighting its potential in Z."
- **Call to action or reflection**
  - Inspire readers to consider the implications of your research or take action based on your findings.
  - Example: "These results underscore the urgency of implementing X strategies to address Y."

- **Use a memorable closing sentence**
  - Craft a statement that resonates with the reader and reinforces the study's importance.
  - Example: "As we look to the future, this research stands as a stepping stone toward a more innovative and sustainable approach to Z."

Tips for crafting a strong ending

- **Be confident but realistic**
  - Avoid overstating your contributions; acknowledge the limitations while highlighting strengths.
  - Example: "While further research is needed, this study lays the groundwork for..."
- **Maintain a positive tone**
  - End with optimism and forward-looking statements.
  - Example: "The insights gained here hold promise for addressing..."
- **Avoid introducing new information**
  - Ensure all key points are addressed earlier in the paper; the conclusion is for synthesis and final thoughts.
- **Be concise**
  - Keep the ending brief but impactful, summarizing the essence of your work without unnecessary elaboration.

Why ending strong matters

- **Leaves a lasting impression:** A well-crafted conclusion stays with the reader long after they finish reading.
- **Enhances credibility:** Demonstrates thoughtful reflection and effective communication.
- **Encourages further engagement:** Motivates readers to explore the topic further or consider practical applications.
- **Brings closure:** Provides a satisfying and logical end to the narrative of the paper.

Ending on a strong note ensures your research resonates with the audience and emphasizes its contribution to the academic or professional field.

3.9 SUMMARY

The chapter "Key Skills for Effective Paper Writing" provides a comprehensive guide to mastering the essential skills required for writing a well-structured and impactful research paper. It covers the critical aspects of each section, including crafting a clear and concise *Introduction*, conducting a thorough and analytical *Literature Review*, and presenting detailed and logically organized *Methods*. The chapter emphasizes the importance of clarity and precision in the *Results*, critical thinking and contextualization in the *Discussion*, and summarization and forward-thinking in the *Conclusions*. By integrating skills such as proper citation, statistical accuracy, logical flow, and effective use of visuals, the chapter equips researchers with the tools to communicate their findings effectively while adhering to academic and professional standards.

REVIEW QUESTIONS

1. What are the main sections of a research paper?
2. What role does clarity play in writing an introduction?
3. How can the use of keywords enhance the relevance of an introduction?
4. Why is balancing creativity and formality important in an introduction?
5. What is the significance of critical thinking in a literature review?
6. How can identifying gaps in the literature improve the quality of research?
7. What is the importance of synthesis in writing a literature review?
8. Why is consistency in terminology essential in the methodology section?
9. How can brevity be maintained without omitting key details in the methods section?
10. What skills are necessary for presenting statistical and analytical information?
11. How does the use of visuals enhance the presentation of results?
12. What is the role of logical organization in reporting results?
13. Why is statistical accuracy crucial in the results section?
14. How does contextualizing results contribute to a strong discussion section?
15. What are the benefits of speculating reasonably in a discussion?
16. Why is identifying future research directions significant?
17. What are the key components of an effective conclusion?
18. How can repetition be avoided when summarizing key findings?
19. Why is it important to end on a strong note in the conclusion?
20. What is the importance of following journal or institutional guidelines when writing a paper?
21. How can proper citation prevent plagiarism and enhance credibility?
22. Why is maintaining a balance between clarity and conciseness critical across all sections of a paper?

Chapter 4

Intellectual property rights and the patenting process

4.1 INTRODUCTION

The realm of Intellectual Property Rights (IPR) has seen unprecedented growth in recent years, reflecting the increasing global emphasis on protecting innovation and creativity. This year's remarkable surge in IPR applications, totalling 568,049, a 7.5% increase from the previous year, underscores this trend. Patent filings have notably risen by 13.57%, with domestic filings making up an impressive 44.41% of the total, a significant jump from 41.58% the previous year. The design domain experienced a remarkable 59.38% growth in applications, complemented by improved examination and disposal rates. Trademarks also saw robust activity, with 447,805 applications filed and a notable reduction in examination pendency to under a month, resulting in a 2.12% increase in registrations and an 8.11% rise in disposals. Meanwhile, the copyright sector received 30,988 applications, leading to 20,673 registrations and 20,820 disposals, demonstrating sustained interest in safeguarding creative works.

This chapter delves into the nature and significance of intellectual property (IP), encompassing patents, designs, trademarks, and copyrights. It explores the processes of patenting and development, highlighting the role of technological research and innovation. Additionally, it examines the international landscape of IP, including cooperation and the procedural intricacies of patent grants and patenting under the Patent Cooperation Treaty (PCT). By understanding these aspects, readers can gain valuable insights into the critical role IP plays in fostering innovation and economic growth.

4.2 NATURE OF INTELLECTUAL PROPERTY RIGHTS (IPR)

IPR encompasses a set of legal rights granted to individuals or entities over creations of the mind, fostering innovation, creativity, and the protection of intangible assets [70]. These rights are classified into patents, trademarks,

DOI: 10.1201/9781003485827-4

copyrights, and trade secrets. Patents safeguard novel inventions, granting exclusive rights to the inventor for a specified duration. Trademarks protect distinctive signs, symbols, or phrases identifying products or services, ensuring brand recognition and consumer trust. Copyrights safeguard original artistic, literary, or musical works, providing creators exclusive rights over their creations. Trade secrets involve confidential business information, granting protection against unauthorized use or disclosure. IPR not only incentivize innovation but also fosters economic growth, encouraging the development and dissemination of new ideas while balancing the rights of creators and the public interest.

IPR encompasses the collection of rights safeguarding intangible assets owned by individuals or entities, shielding them from unauthorized use [71]. Intangible assets pertain to non-physical properties, asserting ownership in various forms of IP. Examples of these rights encompass:

- Patents
- Domain names
- Industrial designs
- Confidential information
- Inventions
- Moral rights
- Database rights
- Creative works by authors
- Service marks
- Logos
- Trademarks
- Design rights
- Business or trade names
- Commercial secrets
- Computer software

IPR protects these diverse assets, granting exclusive rights to creators or owners and regulating their use, reproduction, or commercialization without proper consent.

IPR encompasses several types of legal protections for creations of the mind, aiming to encourage innovation and creativity while safeguarding the rights of creators.

The nature of IP is rooted in its ability to balance individual rights with societal progress. By granting exclusive rights, IP encourages innovation and investment while ensuring that knowledge and creativity ultimately benefit the public. These protections are territorial, varying across jurisdictions, but international frameworks, such as the World Intellectual Property Organization (WIPO), promote global harmonization. Recognizing the value and intricacies of IP is essential in today's knowledge-driven economy.

4.3 TYPES OF INTELLECTUAL PROPERTY RIGHTS (IPR)

IPR encompasses various legal protections that safeguard the creations of the mind, fostering innovation and creativity [72]. The primary types of IPR include:

4.3.1 Patents

A patent is a form of IP that grants the inventor exclusive rights to their invention, preventing others from making, using, selling, or distributing the invention without the inventor's consent for a specific period, usually 20 years from the filing date [73].

Key features

- **Novelty:** The invention must be new and not disclosed to the public before the filing date.
- **Non-obviousness:** The invention must involve an inventive step, meaning it should not be obvious to someone skilled in the field.
- **Industrial applicability:** The invention must have practical utility and be capable of being used in industry.

Types of patents

- **Utility patents:** Protect functional aspects of inventions, such as processes, machines, or compositions of matter.
- **Design patents:** Protect the ornamental or aesthetic design of a product.
- **Plant patents:** Protect new and distinct plant varieties reproduced asexually.

Benefits of patents

- **Incentivizes innovation:** Encourages inventors to develop new technologies by offering legal protection.
- **Monetary gains:** Provides opportunities for licensing, commercialization, and generating revenue.
- **Competitive advantage:** Secures a market position by restricting competitors from using the invention.

Limitations of patents

- **Costly and time-consuming:** Filing, maintaining, and enforcing patents require significant financial and time investments.
- **Limited duration:** Protection is time-bound, after which the invention enters the public domain.

- **Territorial nature:** Patent rights are limited to the jurisdiction where the patent is granted.

Examples of patented innovations

- Alexander Graham Bell's telephone.
- Thomas Edison's light bulb.
- Modern-day pharmaceuticals and technologies like smartphones.

Patenting process

- Conduct a prior art search to ensure novelty.
- Draft and file a patent application with detailed specifications.
- Examination by the patent office.
- Granting of the patent upon meeting all legal requirements.

Patents are crucial in fostering technological advancements and ensuring inventors can reap the benefits of their ingenuity.

4.3.2 Trademarks

A trademark is a recognizable sign, design, word, or expression that identifies and distinguishes the products or services of a particular source from those of others [74]. It represents the brand's identity and assures consumers of the quality and origin of goods or services.

Key features

- **Distinctiveness:** Must be unique and capable of distinguishing the goods or services of one entity from others.
- **Non-deceptiveness:** Should not mislead the public regarding the nature, quality, or origin of the goods or services.
- **Exclusive rights:** Provides the owner with the exclusive right to use the mark and prevent unauthorized use by others.

Types of trademarks

- **Word marks:** Words, letters, or numerals (e.g., "Nike").
- **Design marks:** Logos or symbols (e.g., Apple's logo).
- **Combined marks:** A combination of words and designs.
- **Sound marks:** Distinctive sounds associated with a brand (e.g., the MGM lion's roar).
- **Three-dimensional marks:** Shapes that distinguish goods (e.g., Coca-Cola's bottle design).
- **Colour marks:** Specific colours uniquely identifying a brand (e.g., Tiffany & Co.'s blue).

Benefits of trademarks

- **Brand recognition:** Helps consumers identify and associate products or services with a specific company.
- **Market value:** Enhances brand equity and adds value to the business.
- **Legal protection:** Provides a basis for legal action against infringement.
- **Customer trust:** Signals quality and reliability to consumers.

Limitations of trademarks

- **Renewal requirement:** This must be renewed periodically to maintain protection.
- **Costly enforcement:** Protecting a trademark against infringement can be expensive.
- **Territorial scope:** Protection is jurisdiction-specific unless registered internationally.

Examples of famous trademarks

- McDonald's Golden Arches.
- Nike's "Swoosh" logo.
- Google's wordmark.

Trademark registration process

- Conduct a trademark search to ensure uniqueness.
- File a trademark application with the relevant authority.
- Examination and publication for public objections.
- Granting of trademark upon meeting all legal requirements.

Trademarks play a pivotal role in building brand identity, protecting IP, and fostering consumer trust. They ensure that businesses can differentiate themselves in competitive markets.

4.3.3 Copyrights

Copyright is a form of IP that grants creators of original works exclusive rights to use, reproduce, distribute, and display their creations [75]. It applies to literary, artistic, musical, and certain other intellectual works. Copyright protects the expression of ideas, not the ideas themselves.

Key features

- **Automatic protection:** Copyright arises automatically upon the creation of a work in a tangible form; no registration is required, though registration can strengthen legal claims.

- **Exclusive rights:** Includes rights to reproduce, distribute, perform, display, and create derivative works.
- **Duration:** Generally lasts for the lifetime of the author plus 50 to 70 years, depending on jurisdiction.
- **Territorial:** Protection is jurisdiction-specific but can be extended internationally through treaties.

Types of works covered

- **Literary works:** Books, articles, software, etc.
- **Musical works:** Compositions, lyrics, and recordings.
- **Artistic works:** Paintings, drawings, sculptures, and architecture.
- **Dramatic works:** Plays, scripts, and choreography.
- **Cinematographic works:** Films, videos, and animations.
- **Other intellectual works:** Databases, advertisements, and broadcasts.

Benefits of copyright

- **Economic rights:** Enables creators to monetize their works through licensing and sales.
- **Moral rights:** Protects the integrity and attribution of the creator's work.
- **Legal recourse:** Provides a basis for legal action against unauthorized use.
- **Encourages creativity:** Ensures creators can benefit from their work, incentivizing innovation.

Limitations of copyright

- **Idea vs. expression:** Protects only the expression of an idea, not the idea itself.
- **Fair use:** Certain uses (e.g., criticism, education, and parody) may not require permission.
- **Duration limits:** Copyright eventually expires, after which works enter the public domain.

Examples of copyrighted works

- J.K. Rowling's *Harry Potter* series.
- Beethoven's symphonies.
- Pixar's animated films.

Copyright registration process

- Prepare the work for submission.
- File an application with the relevant copyright office.
- Pay the applicable fees.
- Receive a copyright certificate upon approval.

Infringement and enforcement

- Unauthorized reproduction, distribution, or adaptation constitutes infringement.
- Legal remedies include fines, damages, and injunctions.

International protection

- Governed by treaties like the Berne Convention and the Universal Copyright Convention.
- Provides reciprocal protection among member countries.

Copyrights are essential for protecting the rights of creators, fostering innovation, and ensuring the continued production of cultural and intellectual works.

4.3.4 Design rights

Design rights protect the visual appearance, shape, configuration, pattern, or ornamentation of a product [76]. They ensure exclusive rights to the creator or owner of the design, preventing others from reproducing or imitating it without permission.

Key features

- **Focus on aesthetics:** Protects the appearance of a product rather than its functional aspects.
- **Exclusive rights:** Provides the owner with the ability to prevent unauthorized use of the design.
- **Registration requirement:** Most jurisdictions require formal registration to enforce design rights, though unregistered design rights may offer limited protection in some cases.
- **Duration:** Registered design rights typically last 10–25 years, subject to periodic renewals, while unregistered rights may last for a shorter period (e.g., 3–5 years).

Scope of protection

Design rights cover features such as:

- **Shape and configuration:** Three-dimensional aspects of a product (e.g., the body of a car).
- **Pattern and ornamentation:** Two-dimensional elements (e.g., fabric patterns and wallpaper designs).
- **Combination of features:** The overall aesthetic resulting from multiple design elements.

Benefits of design rights

- **Market differentiation:** Helps establish a unique brand identity through product aesthetics.
- **Commercial value:** Enhances the appeal of a product, boosting sales and revenue.
- **Legal protection:** Enables enforcement against counterfeit or copycat designs.
- **Incentivizes creativity:** Encourages innovation in design industries.

Limitations of design rights

- **Functionality exclusion:** This does not protect the functional aspects of a product; patents are required for functional innovations.
- **Independent creation:** Offers no protection if someone independently creates a similar design without copying.
- **Time-bound protection:** Requires periodic renewals and eventually expires, allowing designs to enter the public domain.

Examples of protected designs

- The iconic shape of the Coca-Cola bottle.
- The sleek design of an Apple iPhone.
- Designer patterns used in Louis Vuitton handbags.

Registration process for design rights

- **Prepare the application:** Include high-quality representations (drawings and photos) of the design.
- **File the application:** Submit to the relevant IP office.
- **Examination:** Authorities check compliance with requirements such as novelty and originality.
- **Grant of rights:** Upon approval, a certificate of registration is issued.

Infringement and enforcement

- Unauthorized reproduction or use of the protected design constitutes infringement.
- Remedies include legal action for damages, injunctions, or the seizure of infringing goods.

International protection

- **Hague system:** Facilitates the international registration of designs in multiple jurisdictions
- **Regional laws:** The European Union offers protection through Community Design Rights, covering all member states.

Design rights are critical in industries like fashion, consumer electronics, and automotive manufacturing, where aesthetics play a pivotal role in market success. They protect creators' investments, enhance product value, and promote innovation in visual design.

4.3.5 Trade secrets

A trade secret refers to confidential business information that provides a competitive advantage to its owner [77]. Unlike patents or copyrights, trade secrets are not publicly disclosed and are protected through secrecy measures.

Key features

- **Confidentiality:** Information must remain secret to retain its status as a trade secret.
- **Economic value:** The information must have commercial value due to its secrecy.
- **Reasonable measures:** The owner must take steps to maintain its confidentiality, such as using non-disclosure agreements (NDAs) and restricted access.

Types of trade secrets

Trade secrets can include:

- **Formulas:** Recipes like the Coca-Cola formula.
- **Processes:** Manufacturing techniques or proprietary algorithms.
- **Designs:** Unique designs not registered as patents.
- **Business Strategies:** Marketing plans, pricing strategies, or supplier lists.
- **Customer Data:** Exclusive customer databases or profiles.

Advantages of trade secrets

1. **No time limit:** Unlike patents, trade secrets can last indefinitely as long as they remain confidential.
2. **Cost-effective:** Avoids the costs and time associated with registration.
3. **Immediate protection:** No need for approval processes or registration.

Limitations of trade secrets:

- **Vulnerability:** Once disclosed, trade secrets lose their protection.
- **No formal rights:** Unlike patents, there is no exclusive right to prevent others from independently developing or discovering the information.
- **Challenging enforcement:** Proving misappropriation in court can be difficult.

Legal protection of trade secrets

Trade secrets are protected under national and international laws, such as:

- **Uniform Trade Secrets Act (UTSA):** In the United States, it provides a framework for trade secret protection.
- **Defend Trade Secrets Act (DTSA):** Offers federal protection in the United States.
- **World Trade Organization (WTO):** Under the TRIPS Agreement, member countries must ensure trade secret protection.

Examples of famous trade secrets

- The recipe for Coca-Cola.
- Google's search algorithm.
- KFC's blend of 11 herbs and spices.

Maintaining trade secrets

- **Non-disclosure agreements (NDAs):** Legally binding agreements to protect confidential information.
- **Access control:** Restrict access to sensitive information to authorized personnel only.
- **Training:** Educate employees on the importance of confidentiality.
- **Monitoring:** Regular audits to ensure compliance with secrecy protocols.

Infringement and remedies

Trade secret theft or misappropriation occurs when:

- Confidential information is disclosed without authorization.
- Information is acquired through improper means, such as hacking or espionage.

Legal remedies

- **Injunctions:** Prevent further use or disclosure.
- **Damages:** Compensation for economic losses.
- **Criminal charges:** In severe cases, trade secret theft can lead to criminal penalties.

Global considerations

Trade secrets are gaining importance in international business, with countries adopting stronger laws to combat theft and misappropriation. Companies often employ global strategies to ensure consistent protection across jurisdictions.

Trade secrets are invaluable for businesses, offering protection for proprietary knowledge without the limitations of formal registration. However, maintaining their confidentiality requires vigilance, robust security measures, and legal safeguards.

4.3.6 Geographical indications (GIs)

A geographical indication (GI) is a sign used on products that have a specific geographical origin and possess qualities, reputation, or characteristics inherently linked to that origin [78]. GIs are typically applied to agricultural products, foodstuffs, wines, spirits, and handicrafts.

Key features

1. **Geographical link:** The product's uniqueness is directly associated with its place of origin.
2. **Collective right:** GIs are not owned by individuals but by communities or associations from the specified region.
3. **Legal protection:** Ensures that only products meeting specific standards can use the GI label.

Examples of GIs

- **Agricultural products:** Darjeeling Tea (India), Champagne (France), and Parmigiano-Reggiano (Italy).
- **Handicrafts:** Pashmina (India) and Murano Glass (Italy).
- **Spirits and wines:** Scotch Whisky (Scotland) and Tequila (Mexico).

Importance of GIs

- **Economic benefits:** GIs enhance the marketability and premium pricing of products.
- **Cultural preservation:** Protects traditional knowledge and heritage associated with the region.
- **Consumer trust:** Guarantees quality and authenticity to consumers.
- **Prevention of misuse:** Safeguards the reputation of regional products from counterfeit or misrepresentation.

Legal protection of GIs

GIs are protected under various national and international frameworks, such as:

- **TRIPS agreement:** The WTO's Trade-Related Aspects of IPR ensure GI protection among member nations.
- **National laws:** Countries have specific laws, such as the Geographical Indications of Goods (Registration and Protection) Act, 1999, in India.

- **EU system:** The European Union offers protection under two categories: Protected Designation of Origin (PDO) and Protected GI (PGI).

Process of GI registration

- **Application submission:** Producers or associations apply for GI status, detailing the product's origin and characteristics.
- **Verification:** Authorities assess the product's link to the geographical region and its unique attributes.
- **Publication and opposition:** The application is published for public scrutiny and objections.
- **Approval:** Upon successful verification, the GI status is granted.

Benefits of GI protection

- **Economic growth:** Encourages local production and employment.
- **Global recognition:** Establishes the product's reputation in international markets.
- **Prevention of unfair practices:** Stops unauthorized use of GI tags, ensuring genuine products reach consumers.

Challenges in GI protection

- **Counterfeiting:** Unauthorized use of GI tags can dilute brand value.
- **Cost of enforcement:** Monitoring and legal enforcement can be expensive.
- **Limited awareness:** Producers and consumers may lack awareness about GI benefits.
- **Global variations:** Differences in GI laws across countries can complicate protection.

Role in sustainable development

GIs contribute to sustainable development by promoting traditional practices, encouraging environmentally friendly production, and fostering economic stability in rural areas.

GIs play a crucial role in protecting and promoting the cultural and economic identity of regions. They provide a valuable tool for enhancing market opportunities while preserving heritage and fostering consumer trust. Ensuring robust legal frameworks and raising awareness about GIs are essential steps for maximizing their potential benefits.

4.3.7 Plant variety protection

Plant Variety Protection (PVP) refers to a legal framework that grants exclusive rights to breeders of new, distinct, uniform, and stable plant varieties [79].

These rights ensure that breeders can control the production, reproduction, and commercialization of their plant varieties for a specified period.

Key features of PVP

- **Breeder's rights:** Breeders can control the use of their plant variety for commercial purposes.
- **Distinctness:** The plant variety must be clearly distinguishable from existing varieties.
- **Uniformity and stability:** The variety must show consistent traits across generations and environments.
- **Novelty:** The plant variety should not have been commercially exploited before applying for protection.

Benefits of PVP

- **Encourages innovation:** Promotes investment in plant breeding and agricultural research.
- **Economic returns:** Breeders benefit from royalties and exclusive commercialization rights.
- **Food security:** Supports the development of high-yield, disease-resistant, and climate-resilient crops.
- **Global trade:** Facilitates international recognition and exchange of plant varieties.

Examples of protected plant varieties

- High-yield rice or wheat varieties.
- Disease-resistant fruits like certain apple or banana strains.
- Ornamental plants with unique aesthetic characteristics.

International framework for PVP

- **UPOV convention:** The International Union for the Protection of New Varieties of Plants (UPOV) establishes guidelines for PVP. Countries adhering to UPOV follow a standardized approach to protect breeders' rights.
- **TRIPS agreement:** The WTO's Trade-Related Aspects of IPR mandates member nations to provide protection for plant varieties, either through patents, PVP systems, or a combination of both.

National implementation of PVP

- **India:** The Protection of Plant Varieties and Farmers' Rights Act, 2001 (PPV&FR Act), provides protection for breeders while recognizing farmers' rights to save, use, and exchange seeds.

- **USA:** The Plant Variety Protection Act (PVPA) protects sexually reproduced and tuber-propagated plant varieties.
- **European Union:** The Community Plant Variety Office (CPVO) grants PVP rights across EU member states.

Process of PVP registration

- **Application submission:** Breeders apply to the national authority with details about the variety.
- **Testing:** The variety undergoes Distinctness, Uniformity, and Stability (DUS) testing.
- **Publication and opposition:** The application is published, allowing objections.
- **Grant of rights:** Upon successful evaluation, PVP rights are granted.

Farmers' rights and PVP

Many national frameworks, like India's PPV&FR Act, balance breeders' rights with farmers' rights, allowing farmers to save and reuse seeds of protected varieties.

Challenges in PVP

- **High costs:** The application and testing process can be expensive for breeders.
- **Limited awareness:** Farmers and small-scale breeders may lack knowledge about PVP.
- **Ethical concerns:** Critics argue that PVP may restrict farmers' traditional seed-saving practices.
- **Biopiracy risks:** Protection systems must ensure that traditional and indigenous varieties are not exploited unfairly.

Role of PVP in sustainable agriculture

- **Development of resilient crops:** Encourages the breeding of varieties that withstand environmental challenges.
- **Preservation of biodiversity:** Ensures the conservation of traditional plant varieties.
- **Support for small breeders:** Provides recognition and economic incentives to local and small-scale breeders.

PVP is a critical component of IPR in agriculture. By balancing breeders' innovation incentives with farmers' traditional practices, PVP fosters agricultural advancement, economic growth, and global food security. Robust legal frameworks and awareness campaigns are essential to ensure the equitable implementation of PVP systems worldwide.

4.3.8 Integrated circuit layout design rights

Integrated Circuit Layout Design Rights (ICLDR) refer to the legal protection granted to the unique design and arrangement of electronic circuits in an integrated circuit (IC) [80]. These rights ensure that the creators of IC layouts have exclusive control over the reproduction, distribution, and commercialization of their designs.

Key features of ICLDR

- **Originality:** The layout design must be original and a result of the creator's intellectual effort.
- **Fixed form:** The design must be embodied in a material form, such as a chip or mask.
- **Exclusivity:** Protection grants exclusive rights to the creator, preventing unauthorized reproduction or commercial use.
- **Non-triviality:** The layout must not be commonplace among designers and manufacturers.

Scope of protection

- **Protected elements:** Includes the three-dimensional arrangement of transistors, resistors, capacitors, and other circuit components.
- **Excluded elements:** The functional aspects or ideas behind the design, as well as designs that are standard or commonplace.

Importance of ICLDR

- **Encourages innovation:** Provides incentives for designing new and efficient IC layouts.
- **Supports the electronics industry:** Protects investments in research and development of microelectronics.
- **Prevents copying:** Discourages unauthorized duplication and counterfeiting of IC designs.

International framework

- **TRIPS agreement:** Mandates World Trade Organization (WTO) members to provide protection for IC layout designs.
- **Washington treaty (1989):** The Treaty on Intellectual Property in Respect of Integrated Circuits establishes guidelines for protecting IC layouts.

National implementation

- **India:** The Semiconductor Integrated Circuits Layout-Design Act, 2000, provides exclusive rights for IC layouts.

- **United States:** The Semiconductor Chip Protection Act (SCPA) of 1984 offers a 10-year protection for original IC designs.
- **European Union:** Protects IC layout designs under various national laws aligned with the TRIPS Agreement.

Rights conferred by ICLDR

- **Reproduction rights:** The creator has the sole right to reproduce the layout design.
- **Commercial rights:** Exclusive rights to import, sell, or distribute ICs made using the protected layout.
- **Exemptions:** Reverse engineering for educational or research purposes is generally allowed.

Duration of protection

- The protection typically lasts for 10 years from the date of registration or the first commercial exploitation, depending on the jurisdiction.

Examples of protected layout designs

- Advanced microprocessor chips used in computers.
- Custom IC designs for smartphones or IoT devices.
- Specialized IC layouts for medical or aerospace applications.

Challenges in ICLDR

- **Complexity of enforcement:** Identifying and proving infringement can be technically challenging.
- **Short lifecycle:** The rapid evolution of technology may outpace the protection period.
- **High costs:** Designing and registering IC layouts can be expensive for small firms.

Benefits of ICLDR

- **Promotes technological progress:** Encourages companies to invest in innovative designs.
- **Boosts economic growth:** Protects the competitive edge of technology-driven industries.
- **Global standardization:** Aligns national laws with international practices, facilitating global trade.

ICLDRs are crucial for protecting IP in the microelectronics industry. By safeguarding the intricate designs of IC layouts, these rights ensure that innovators receive recognition and economic benefits while fostering innovation and competition in the rapidly evolving technology sector. Strengthening enforcement mechanisms and raising awareness are essential to maximize the benefits of ICLDR.

4.4 PROCESS OF PATENTING

The patenting process is a structured procedure that enables inventors to secure exclusive rights for their inventions [81]. It ensures that innovative ideas are legally protected, preventing unauthorized use or replication while promoting technological advancement. The process involves several key stages:

4.4.1 Identifying the invention

Identifying the invention is the foundational step in the patenting process. This stage involves defining and documenting the unique aspects of an idea or innovation to ensure it meets the criteria for patentability. A clear understanding of the invention is crucial to demonstrate its originality, utility, and industrial applicability.

Key elements of identifying the invention

- **Define the problem and solution**
  - Articulate the problem your invention addresses.
  - Highlight how your solution is innovative compared to existing alternatives.
- **Establish novelty**
  - Ensure the invention is entirely new and has not been disclosed in any public domain, including patents, publications, or prior use.
- **Utility and applicability**
  - Confirm the invention serves a useful purpose and can be applied practically in a specific field or industry.
- **Non-obviousness**
  - Verify that the invention is not an obvious improvement or variation of existing technologies to someone skilled in the relevant field.
- **Comprehensive documentation**
  - Record every detail of the invention, including:
    1. Diagrams and sketches.
    2. Functional descriptions and technical specifications.
    3. Potential applications and benefits.
- **Search for similar inventions**
  - Conduct preliminary research using patent databases to identify any existing inventions that might overlap or conflict with your idea.

- **Categorize the Invention**
  - Determine the type of IP protection needed (e.g., utility patent, design patent, or plant patent).

Significance in the patenting process

Identifying the invention ensures clarity and direction for the subsequent stages of patenting. It establishes a solid foundation for drafting a robust patent application and avoids potential disputes or rejections during examination. A well-defined invention maximizes the chances of obtaining patent protection and leveraging its commercial potential.

4.4.2 Conducting a patent search

A patent search is a critical step in the patenting process that involves investigating existing patents, published applications, and other public disclosures to determine if an invention is novel and patentable [82]. This step minimizes the risk of infringement and ensures the invention meets the criteria for patent protection.

Key steps in conducting a patent search

- **Define the scope of the search**
  - Identify the key features, components, and applications of your invention.
  - Determine relevant keywords, technical terms, and classifications for an effective search.
- **Use patent classification systems**
  - Leverage classification systems such as:
    1. International Patent Classification (IPC).
    2. Cooperative Patent Classification (CPC).
    3. US Patent Classification (USPC).
  - These systems help narrow down search results to relevant technology areas.
- **Search patent databases**
  - Explore major databases like:
    1. WIPO.
    2. United States Patent and Trademark Office (USPTO).
    3. European Patent Office (EPO).
    4. National databases for regional patents.
- **Analyze prior art**
  - Examine existing patents, applications, and non-patent literature (NPL) to identify similarities.
  - Focus on the claims, descriptions, and drawings to assess overlap with your invention.

- **Evaluate patentability criteria**
  - Confirm the invention's novelty, non-obviousness, and utility compared to prior art.
  - Ensure it does not fall under excluded categories for patent protection.
- **Document search results**
  - Maintain detailed records of search terms, databases used, and results.
  - Note any potential conflicts or similarities that may require adjustments to your invention or application.
- **Seek professional assistance**
  - Engage a patent attorney or search professional for complex inventions or comprehensive searches.

Significance of a patent search

Conducting a patent search reduces the risk of investing time and resources in a non-patentable invention. It helps refine the invention by identifying gaps or areas for improvement and avoids potential infringement issues. This proactive approach strengthens the patent application and ensures compliance with legal requirements.

By thoroughly exploring the patent landscape, inventors can position their innovations strategically, enhancing the likelihood of obtaining strong IPR.

4.4.3 Drafting the patent application

Drafting a patent application is a meticulous process that transforms an invention into a legally recognized IP asset. A well-crafted patent application not only describes the invention but also secures its protection by satisfying the legal and technical requirements of the patent office [83].

Key components of a patent application

- Title
  - Concise and descriptive, summarizing the invention's purpose or function.
  - Abstract
  - A brief summary (150–250 words) highlighting the invention's key features, technical field, and primary utility.
  - Written to provide a quick understanding without detailed technical jargon.
- **Background of the invention**
  - Describes the problem or need that the invention addresses.
  - Provides context by explaining existing solutions and their limitations.

- **Summary of the invention**
  - Highlights the novel aspects of the invention.
  - Offers an overview of how the invention solves the stated problem.
- **Detailed description**
  - Provides an in-depth explanation of the invention, including its components, structure, and operation.
  - Written clearly and comprehensively to enable someone skilled in the art to reproduce the invention.
- **Claims**
  - Define the scope of protection sought.
  - Includes independent claims (broad) and dependent claims (narrow and referencing independent claims).
  - Must be precise and legally defensible.
- **Drawings (if applicable)**
  - Visual representations such as diagrams, flowcharts, or schematics to complement the description.
  - Clearly labelled and referenced in the text.
- **Examples and embodiments**
  - Real-world examples or specific embodiments of the invention to demonstrate its functionality and applications.
- **Industrial applicability**
  - Describes how the invention can be applied in industry or practical use cases.
  - References and Prior Art
  - Acknowledge relevant prior art and explain how the invention improves upon or differs from it.

Key skills for drafting a patent application

- **Technical expertise**
  - A deep understanding of the invention and its technical field to describe it accurately.
- **Legal knowledge**
  - Familiarity with patent laws, including novelty, non-obviousness, and utility requirements.
- **Clarity and precision**
  - Avoiding vague language to ensure the invention is comprehensible and defensible.
- **Strategic claim drafting**
  - Writing claims to maximize protection while avoiding unnecessary limitations.
- **Attention to detail**
  - Ensuring all components and aspects of the invention are included and accurately described.

Importance of a well-drafted patent application

A well-drafted patent application strengthens the likelihood of obtaining patent protection and provides a robust foundation for defending the invention against infringement. By clearly defining the invention's scope and emphasizing its uniqueness, the application becomes a valuable asset in securing IPR. Engaging professionals, such as patent attorneys, can further enhance the application's quality and compliance with legal standards.

4.4.4 Filing the patent application

Filing a patent application is the critical step that initiates the process of securing legal protection for an invention [84]. It involves submitting the prepared application to the relevant patent office, ensuring compliance with procedural requirements, and establishing the filing date, which is essential for determining priority rights.

Steps involved in filing a patent application

- **Choose the type of application**
  - **Provisional application:** Allows inventors to establish an early filing date with minimal formalities.
  - **Non-provisional application:** A complete application that undergoes examination and leads to the grant of a patent.
  - **International application (PCT):** Facilitates patent protection in multiple countries under the PCT.
- **Determine the patent office**
  - File with the appropriate patent office based on the inventor's location or target markets (e.g., USPTO, EPO, or national offices).
- **Prepare required documents**
  - Completed application form.
  - Detailed patent specification, including claims and drawings.
  - An abstract summarizing the invention.
  - Declarations, power of attorney (if applicable), and any priority documents.
- **Pay filing fees**
  - Submit the prescribed fees, which may vary based on the type of application, applicant category (e.g., individual, small entity, or large entity), and jurisdiction.
- **Submit the application**
  - File electronically via the patent office's online portal or physically at the designated office.
  - Retain the acknowledgement receipt, which confirms the filing date and application number.

- **Establish priority (if applicable)**
  - If claiming priority from an earlier application, submit the priority documents within the prescribed timeframe.
- **Verify compliance**
  - Ensure that all forms are correctly filled and that the application complies with the legal and procedural requirements of the jurisdiction.

Key considerations for filing

- **Timeliness**
  - File as early as possible to establish priority rights and prevent others from filing a similar invention.
- **Jurisdictional rules**
  - Adhere to the specific filing requirements of the patent office where the application is being submitted.
- **Confidentiality**
  - Maintain the confidentiality of the invention until the application is filed to preserve its novelty.
- **Engaging professionals**
  - Consult patent attorneys or agents to ensure the application is accurate and comprehensive.

Post-filing process

- **Acknowledgement and application number**
  - The patent office provides an application number and confirms the filing date.
- **Publication**
  - Most patent applications are published 18 months after the filing date or priority date unless a non-publication request is made.
- **Examination**
  - The patent office examines the application to assess compliance with patentability criteria (novelty, non-obviousness, and industrial applicability).
- **Response to office actions**
  - Address objections or rejections raised by the examiner during the examination phase.

Filing a patent application correctly is essential to secure IPR and initiate the examination process. Proper preparation, compliance with legal requirements, and timely submission can significantly enhance the chances of a successful outcome.

4.4.5 Patent examination

Patent examination is a critical phase in the patenting process, where the patent office evaluates the submitted application to determine if the invention meets the criteria for patentability [85]. This process ensures that only inventions that are novel, non-obvious, and industrially applicable receive legal protection.

Key stages of patent examination

- **Formal examination**
    - The patent office verifies that the application adheres to procedural requirements, including the completeness of documents, correct formatting, and payment of fees.
    - Missing or incorrect information may result in formal objections, which must be addressed promptly.
- **Publication**
    - The application is published, typically 18 months after the filing or priority date, making the invention details available to the public.
- **Substantive examination**
    - The core of the patent examination process, where the examiner assesses:
        1. **Novelty:** Whether the invention is new and has not been disclosed in prior art.
        2. **Non-obviousness:** Whether the invention represents a significant step beyond existing knowledge and would not be obvious to a skilled person in the field.
        3. **Industrial applicability:** Whether the invention can be practically applied in an industry.
- **Search report and office actions**
    - The examiner conducts a prior art search and issues a search report highlighting similar existing inventions.
    - Office actions outline objections or rejections, requiring the applicant to respond with clarifications, amendments, or arguments.
- **Amendments and responses**
    - The applicant may modify claims, descriptions, or drawings to address objections raised by the examiner.
    - Clear and precise responses are crucial to overcoming rejections.
- **Grant or rejection**
    - If the examiner is satisfied that the invention meets all patentability criteria, the patent is granted.
    - If objections persist, the application may be rejected, with an option for appeal or review.

Examination criteria

- **Patentable subject matter**
    - The invention must fall within the categories of patentable subject matter defined by the jurisdiction.

- **Novelty**
  - The invention must not have been disclosed in any public document, patent, or use prior to the filing date.
- **Non-obviousness**
  - The invention should involve an inventive step that is not obvious to someone skilled in the relevant field.
- **Utility**
  - The invention must have a practical application and be capable of industrial use.

Strategies for a successful patent examination

- **Thorough prior art search**
  - Conduct a comprehensive search before filing to identify potential conflicts and refine claims accordingly.
- **Clear and precise claims**
  - Ensure that claims are specific, well-defined, and adequately supported by the description.
- **Effective communication with examiners**
  - Provide detailed, logical, and evidence-backed responses to office actions.
- **Engaging professionals**
  - Collaborate with patent attorneys or agents to navigate complex examination processes effectively.

Post-examination Steps

- **Grant of patent**
  - Upon successful examination, the patent is granted, providing exclusive rights to the inventor.
- **Publication of grant**
  - The granted patent is published, making the details accessible to the public.
- **Maintenance**
  - Periodic fees must be paid to maintain the patent's validity.

Patent examination is a meticulous process that ensures the integrity and value of granted patents. By meeting the required standards and addressing examiner feedback effectively, applicants can secure robust protection for their inventions.

4.4.6 Publication of the application

The publication of a patent application is a crucial step in the patenting process, providing transparency and allowing the public to access the details of the invention. This stage ensures that the innovation is publicly disclosed while the inventor retains the potential for exclusive rights.

Key aspects of application publication

- **Timing**
  - Patent applications are typically published **18 months** after the filing date or priority date, whichever is earlier.
  - In some cases, applicants may request **early publication** to expedite the process.
- **Content of the publication**
  - The publication includes:
    1. Title of the invention.
    2. An abstract summarizing the invention.
    3. Detailed description, including claims and drawings.
    4. Applicant and inventor details.
- **Public access**
  - Once published, the application becomes part of the **public domain**, allowing anyone to view its details through patent office databases.
- **Rights of the applicant**
  - Publication grants the applicant **provisional rights**, enabling them to seek reasonable royalties from parties that use the invention after publication but before the patent is granted.

Significance of publication

- **Disclosure to the public**
  - Promotes the dissemination of knowledge, encouraging further innovation and collaboration.
- **Establishes prior art**
  - The publication acts as evidence of the invention's existence and filing date, preventing others from claiming it.
- **Facilitates examination**
  - Public access allows third parties to submit prior art references, aiding the patent office in assessing novelty and non-obviousness.
- **Market awareness**
  - Companies and investors can evaluate the invention's potential, opening opportunities for partnerships, licensing, or commercialization.

The concept of **Early Publication** vs. **Regular Publication** typically refers to the timing and the nature of when a patent application is made publicly available. Here's an explanation of these two types of publication.

1. **Early publication**
   Early Publication refers to the process of making a patent application available to the public before the standard publication date, which is typically 18 months after the filing date or the priority date. This early disclosure is often voluntary and can be requested by the applicant.

Key features of early publication

- **Timing:** Early publication occurs before the usual 18-month period. It may happen shortly after the filing date, depending on the jurisdiction.
- **Applicant's choice:** The applicant can request early publication, often to expedite the process or to signal to the market that they are seeking protection for the invention.
- **Benefits**
  - It can help establish the applicant's priority over competitors.
  - It can be used to deter potential infringers by publicly disclosing the application.
  - It may help in the process of licensing or partnering by showing others that a patent is pending.
- **Drawbacks**
  - Early publication means the details of the invention are made public before the patent is granted, which could allow competitors to design around the invention.
  - If the patent is later rejected, the applicant may have disclosed an idea that they cannot protect.

2. **Regular publication**
   Regular publication refers to the standard process where the patent application is made publicly available 18 months after the filing date or priority date. This is the default method of publication unless the applicant specifically requests early publication or a different arrangement.

Key features of regular publication

- **Timing:** Occurs 18 months after the filing date or priority date, as per standard patent laws.
- **Automatic process:** This is the default procedure; applicants do not need to request regular publication.
- **Benefits**
  - The invention is kept confidential for a longer period, allowing the applicant to assess the commercial viability and the patentability of the invention before it becomes public.
  - It provides the applicant more time to decide whether to pursue the patent or abandon it.
  - If the patent is not granted, the applicant's idea remains confidential.
- **Drawbacks**
  - The applicant's invention is not disclosed to the public for 18 months, which may delay the benefits of establishing priority and deterring competitors.
  - Competitors may independently invent similar technologies, potentially leading to conflicting patent applications (Table 4.1).

Table 4.1 Comparison between early and regular publication

Aspect	*Early publication*	*Regular publication*
Timing of Publication	Before 18 months (usually shortly after filing)	18 months after the filing or priority date
Applicant's Control	Applicant must request it	Automatic unless requested otherwise
Advantages	Establishes early priority	More time to evaluate commercial potential
	Deters competitors early	Keeps invention confidential for longer
	Useful for licensing and partnerships	Protects against public disclosure before the patent is granted
Disadvantages	Invention details are made public early	Competitors may file similar inventions before publication
	Risk if the patent is not granted or is abandoned	Delay in deterring competitors
Patent Pending Status	Early indication of pending status	Patent pending only after regular publication

Implications of publication

- **Provisional rights**
  - Enables inventors to seek compensation for unauthorized use after publication.
- **Strategic positioning**
  - Helps in assessing competitors and gauging market interest.
- **Loss of confidentiality**
  - Once published, the details are no longer confidential, emphasizing the importance of thorough preparation before filing.

Strategies for effective publication

- **Ensure accuracy**
  - Double-check claims, descriptions, and drawings to avoid errors or ambiguities.
- **Leverage early publication**
  - Use early publication strategically to secure rights or attract potential collaborators.
- **Monitor published applications**
  - Regularly review patent office publications to track competing innovations.

Publication of the application is a pivotal stage in the patent process, balancing the inventor's rights with the public's access to technological advancements. Proper planning and strategic use of this phase can significantly enhance the value and impact of the invention.

4.4.7 Grant of patent

The grant of a patent is the culmination of the patenting process, conferring exclusive rights to the inventor or applicant for a specific period, typically 20 years from the filing date [86]. This stage signifies that the invention has successfully met all the legal and technical requirements of patentability.

Steps leading to the grant of a patent

- **Completion of examination**
  - After the **examination** stage, the patent office issues an **examination report** outlining the application's compliance with patentability criteria (novelty, non-obviousness, and industrial applicability).
- **Response to objections**
  - If the examination report includes objections, the applicant must respond and make necessary amendments to the application to address the issues raised.
- **Pre-grant opposition**
  - During the opposition period, third parties may challenge the application, providing evidence that the invention does not meet patentability criteria.
- **Decision by the patent office**
  - If all objections are resolved and no opposition is successful, the patent office issues a **grant order** confirming the patent.

Key features of a granted patent

- **Exclusive rights**
  - The patent holder gains the exclusive right to make, use, sell, or license the invention within the jurisdiction of the granting patent office.
- **Territorial nature**
  - A patent is only enforceable in the country or region where it is granted.
- **Public disclosure**
  - The details of the invention are made publicly accessible in exchange for granting exclusive rights.
- **Maintenance requirements**
  - The patent holder must pay periodic renewal fees to keep the patent active throughout its term.

Significance of a patent grant

- **Legal protection**
  - Prevents unauthorized use, reproduction, or sale of the patented invention.
- **Commercial advantage**
  - Provides a competitive edge by allowing the patent holder to monopolize the market for the invention.
- **Monetization opportunities**
  - Enables licensing agreements, partnerships, and royalties, offering financial benefits.
- **Encouragement for innovation**
  - Protecting IP incentivizes further research and development.

Post-grant activities

- **Post-grant opposition**
  - Some jurisdictions allow third parties to challenge the validity of the granted patent within a specified period.
- **Enforcement of rights**
  - Patent holders can take legal action against infringement to protect their rights.
- **Commercial exploitation**
  - The patent holder can license the technology, sell the patent, or use it to establish a market presence.

The grant of a patent represents a critical milestone in safeguarding IP. It provides inventors with exclusive rights, enabling them to capitalize on their innovations while contributing to technological progress. However, obtaining and maintaining a patent requires careful planning, adherence to legal requirements, and strategic management to maximize its benefits.

4.4.8 Maintenance of the patent

Once a patent is granted, it requires regular upkeep to ensure that the exclusive rights remain in force throughout its term [87]. This process, known as patent maintenance, involves the payment of renewal or maintenance fees at specified intervals.

Key aspects of patent maintenance

- **Renewal fees**
  - Patent holders must pay annual renewal fees to keep the patent valid.
  - The fees typically increase progressively over the patent's term to encourage commercialization and discourage the retention of non-utilized patents.

- **Grace period**
  - Many jurisdictions allow a grace period after the due date for payment of renewal fees, during which the patent can be maintained by paying additional late fees.
- **Lapse of patent**
  - Failure to pay renewal fees results in the patent lapsing, making the invention freely available for public use.

Purpose of patent maintenance

- **Incentivizing commercialization**
  - Increasing fees ensures that only economically valuable patents are maintained, promoting the active use of IP.
- **Freeing resources**
  - Non-maintained patents enter the public domain, fostering further innovation and use by others.

Factors to consider for patent maintenance

- **Commercial viability**
  - Patent holders should evaluate whether the invention generates sufficient revenue to justify maintenance costs.
- **Geographical scope**
  - Maintaining patents in multiple jurisdictions can be costly, so strategic decisions should be made about which regions to prioritize.
- **Portfolio management**
  - Companies with extensive patent portfolios often conduct regular audits to decide which patents to maintain, sell, or allow to lapse.

Maintenance timelines and procedures

- **National patents**
  - Each country has specific timelines and fee structures for patent renewal, which must be adhered to by the patent holder.
- **International patents (PCT)**
  - Under the PCT, renewal fees are payable in each country where the patent is granted.
- **Documentation**
  - Patent holders must keep accurate records of due dates and payments to avoid unintentional lapses.

Consequences of non-maintenance

- **Loss of rights**
  - If a patent lapses, the holder loses exclusive rights to the invention.
  - Competitors or other entities can freely use the invention.

- **Impact on licensing**
  - Lapsed patents cannot be licensed or used as assets, potentially reducing revenue opportunities.

Strategies for effective patent maintenance

- **Automated systems**
  - Use patent management software to track due dates and ensure timely payments.
- **Portfolio prioritization**
  - Focus resources on maintaining patents with high commercial or strategic value.
- **Collaboration with experts**
  - Engage patent attorneys or agents to handle maintenance activities, especially for global portfolios.

Patent maintenance is a critical aspect of managing IPR. It ensures that the exclusive rights conferred by a patent remain in force, allowing inventors and organizations to protect and monetize their innovations. By strategically managing renewal fees and prioritizing valuable patents, holders can maximize the benefits of their IP while minimizing costs.

4.5 CHALLENGES IN PATENTING

The process of patenting is intricate and often fraught with challenges that inventors and organizations must navigate to secure IPR [88]. These challenges can arise at various stages, from the initial conception of the invention to the maintenance of the granted patent.

4.5.1 Identifying novelty and inventiveness

Identifying **novelty** and **inventiveness** is crucial in determining whether an invention qualifies for patent protection. Here's a breakdown of both concepts:

Novelty (newness)

- **Definition:** An invention is considered novel if it has not been disclosed in any form before the filing date of the patent application. This means the invention must not be part of the prior art.
- **Challenges in identifying novelty**
  - **Prior art search:** The first step is conducting an exhaustive search of prior art, including patents, publications, product designs, and even oral disclosures. A single prior disclosure can destroy the novelty of an invention.

- **Global perspective:** Novelty is judged globally, not just in the jurisdiction where the patent is filed. Even a foreign publication or product can affect the novelty.
- **Subtle differences:** Sometimes, inventions that appear novel might only be marginally different from prior art. Identifying whether these differences are substantial enough to be considered novel can be tricky.

Inventiveness (Non-obviousness)

- **Definition:** Inventiveness requires that the invention must not be obvious to someone skilled in the relevant field based on the prior art. This is assessed through a subjective evaluation of the invention's inventive step or non-obviousness.
- **Challenges in identifying inventiveness**
  - **Obviousness Test:** The challenge lies in determining whether the invention is a non-obvious improvement over existing knowledge. If an invention is an obvious solution to a problem based on prior art, it cannot be patented.
  - **Different standards:** What is considered "obvious" can vary by jurisdiction and patent examiner, making it difficult to predict outcomes.
  - **Combination of known elements:** In many cases, an invention combines existing elements or ideas in a new way. Whether this combination is obvious or inventive depends on whether the skilled person would have been able to make the leap.
  - **Technological evolution:** In rapidly evolving fields, what is considered inventive can change quickly, making it challenging to assess whether a solution is novel and non-obvious.

Key factors for identifying novelty and inventiveness

- **Prior art review:** Comprehensive analysis of existing patents, scientific literature, and publicly available technologies.
- **Problem-solution approach:** Understanding the problem the invention addresses and how it differs from existing solutions is key to assessing inventiveness.
- **Technical field:** The level of skill and knowledge in the relevant technical field influences the determination of inventiveness.
- **Unexpected results:** If an invention produces unexpected or surprising results, it is more likely to be considered inventive.

Overall, determining novelty and inventiveness involves both objective searches of prior art and subjective evaluations of how an invention stands in relation to what is already known. These assessments are essential for securing patent protection.

4.5.2 Complexity of patent law

The complexity of patent law arises from several factors that make it a multifaceted and intricate area of legal practice [89]. Some key aspects that contribute to this complexity include:

Global nature of patents

- **Jurisdictional variations:** Patent laws differ from country to country, with each jurisdiction having its own rules regarding patentability, the application process, and enforcement. A patent granted in one country does not automatically apply in another, and inventors must often file separate applications in each jurisdiction.
- **International treaties:** While treaties like the PCT aim to streamline the process for international patent protection, they still require navigating through multiple national or regional patent offices, each with its own legal nuances.

Patentability requirements

- **Novelty, inventiveness, and utility:** The basic criteria for patentability—novelty, inventiveness (non-obviousness), and utility (usefulness)—are subjective and often difficult to assess, particularly in emerging technologies. Determining what constitutes prior art, how inventive an improvement is, or whether an invention is useful enough can be complex.
- **Excluded subject matter:** Patent law often excludes certain subject matters, such as abstract ideas, natural phenomena, and laws of nature, from being patentable. These exclusions can be difficult to interpret and vary depending on the jurisdiction.

Detailed legal and technical language

- **Complex terminology:** Patent applications require precise technical and legal language to describe the invention. A poorly written application or ambiguous claims can lead to disputes over interpretation or rejection of the application.
- **Claim construction:** Patent claims define the scope of protection, and how they are written significantly affects the outcome of both patent prosecution and litigation. Determining the proper scope of claims can be a complex legal issue, especially when terms are open to interpretation.

Patent prosecution process

- **Examination and office actions:** Patent applications undergo a thorough examination process where patent examiners review the invention against prior art and assess its patentability. This can involve multiple rounds of office actions, where the applicant may need to amend the claims or argue their case.
- **Delays and costs:** The patent prosecution process can be lengthy and expensive, especially in high-tech or complex fields. Legal fees, patent search costs, and the cost of responding to office actions can add up quickly.

Patent infringement and litigation

- **Infringement analysis:** Determining whether a product or process infringes on an existing patent requires a detailed analysis of the patent claims and the allegedly infringing product. This often involves both legal and technical expertise.
- **Patent litigation:** Patent disputes can lead to lengthy and costly litigation, often involving expert witnesses, complex legal arguments, and strategic considerations about jurisdiction, enforcement, and settlement. The outcome can be unpredictable, particularly in areas with rapidly evolving technologies.
- **Patent trolls:** Some entities, known as "patent trolls," acquire patents solely for the purpose of filing lawsuits against others, often without intending to produce or use the patented technology. This has added another layer of complexity to patent law.

Changes in patent law

- **Evolving legal precedents:** Patent law is dynamic, with new case law emerging frequently, which can affect interpretations of existing laws. Courts may reinterpret patentability standards, and legislative bodies may amend patent laws to address emerging technologies and global trends.
- **Technology-specific challenges:** Rapid advancements in fields like artificial intelligence (AI), biotechnology, and software have raised new questions regarding patent eligibility and the scope of protection. For instance, issues related to patenting algorithms, genetic modifications, or AI-generated inventions remain unsettled in many jurisdictions.

Patent strategies and management

- **Portfolio management:** Companies with extensive patent portfolios must develop strategies to manage their patents, such as deciding which patents to enforce, license, or abandon. Patent portfolios are also valuable assets in mergers and acquisitions, which adds to the complexity of managing IP.

- **Licensing and royalties:** Patent owners often license their patents to others, creating complex contractual agreements that govern the use of the technology. Negotiating licensing terms, calculating royalties, and enforcing license agreements add further complexity.

Ethical and policy issues

- **Access to innovation:** There is ongoing debate about the role of patents in fostering innovation versus creating barriers to access, especially in critical fields like healthcare. Issues such as patenting essential medicines or life-saving technologies raise ethical concerns about the balance between incentivizing innovation and ensuring public access.
- **Patent thickets and antitrust concerns:** In certain industries, patents can accumulate in a way that creates "patent thickets," where multiple overlapping patents make it difficult for others to innovate without infringing on existing patents. This can lead to anti-competitive practices and antitrust issues.

Patent law is complex due to its intersection of technical, legal, and business considerations. Navigating the patent system requires an understanding of both the legal framework and the technical aspects of the invention. As technologies evolve, patent law must also adapt, creating an ongoing challenge for inventors, businesses, and legal professionals.

4.5.3 High costs

The high costs associated with patenting are one of the significant challenges inventors and companies face. These costs can arise at various stages of the patent process, from filing to enforcement, and can vary depending on factors such as the technology involved, the jurisdiction, and the complexity of the invention. Here's an overview of the major cost components:

- **Patent filing fees**
  - **Initial filing fees**
    When filing a patent application, inventors must pay fees to the relevant patent office. These fees can vary depending on the jurisdiction. For example, filing a patent application with the US Patent and Trademark Office (USPTO) can cost anywhere from a few hundred to several thousand dollars, depending on whether the application is provisional or non-provisional, and whether it's filed by an individual, small entity, or large entity.

- **Additional filing costs**
  If an international patent is sought, filing under the PCT involves additional fees, which can add thousands of dollars for each country or region in which protection is sought.
- **Patent search and examination fees**
  - **Patent search**
    Before filing a patent application, it is recommended to conduct a patent search to identify existing patents or prior art that may affect the novelty and inventiveness of the invention. This search is typically done by a professional patent search firm, and costs can range from a few hundred to several thousand dollars.
  - **Examination fees**
    After filing, most jurisdictions require payment for the examination of the patent application. These fees are typically charged by the patent office and may be substantial, depending on the jurisdiction and the complexity of the invention.
- **Attorney and professional fees**
  - **Patent attorneys**
    Patent applications require specialized legal expertise to draft and prosecute. Hiring a patent attorney or agent can be costly, especially for complex technologies. The cost of hiring a patent attorney can range from $5,000 to $15,000 or more for a single patent application, depending on the complexity of the invention and the experience of the attorney.
  - **Additional legal services**
    If a patent is opposed, litigated, or challenged, legal fees can increase significantly. Litigation costs, including court fees, expert witness fees, and attorney fees, can run into the hundreds of thousands or even millions of dollars.
- **Maintenance fees**
  - **Patent maintenance**
    Once a patent is granted, it is necessary to pay maintenance fees at regular intervals (e.g., every 3–5 years) to keep the patent in force. These fees can increase over time and may range from hundreds to thousands of dollars, depending on the jurisdiction and the age of the patent.
  - **Multiple jurisdictions**
    For patents filed in multiple countries, maintenance fees must be paid in each jurisdiction, which can significantly increase the overall cost of maintaining international patent protection.
- **International patent protection**
  - **Translation and filing costs**
    If an inventor seeks protection in multiple countries, they must file separate applications in each jurisdiction or region. This

often involves additional translation costs, as patent documents must be translated into the official languages of each jurisdiction. These translation costs can add up to several thousand dollars per country.

- **Regional patents**
  In regions like Europe, where a single patent application can lead to protection across multiple countries (via the European Patent Office), there are still significant fees for filing, examination, and validation in each country, further increasing the cost of international protection.

- **Patent litigation costs**
  - **Infringement enforcement**
    If a patent holder believes their patent is being infringed, enforcing the patent through litigation can be very costly. Legal fees, court fees, and expert testimony can amount to significant sums, with some cases reaching millions of dollars, especially in industries like pharmaceuticals or technology.
  - **Defending against infringement**
    Conversely, defending against patent infringement allegations can also be expensive, as it may involve counterclaims, expert analysis, and legal defenses. Companies often face the risk of high costs in both defending and enforcing patents.
- **Cost of patent strategy and portfolio management**
  - **Patent portfolio management**
    Companies with large portfolios of patents need to manage them effectively, which can involve significant costs for maintaining, licensing, and litigating patents. This can require dedicated legal and administrative staff or external consultants.
  - **Licensing and royalties**
    Licensing patents to other companies or negotiating royalty agreements involves additional legal and financial resources. These costs may include drafting licensing agreements, monitoring compliance, and resolving disputes.
- **Opportunity costs**
  - **Time and resources**
    The patenting process is lengthy, often taking several years from filing to grant. During this time, inventors and companies must invest significant time and resources in the patent application process, which can delay other aspects of product development or business operations.
  - **Risk of rejection**
    Not all patent applications are granted. If a patent application is rejected, the inventor may have spent considerable resources on a non-patentable invention, resulting in a loss of time and money.

The high costs of patenting can be a barrier to entry for individual inventors, small businesses, and even larger companies, especially in industries with complex technologies. Balancing the cost of obtaining and maintaining patents with the potential benefits of IP protection is a critical consideration for any inventor or business.

4.6 BENEFITS OF PATENTING

Patenting offers a range of benefits for inventors, businesses, and organizations, making it a valuable tool in protecting IP and promoting innovation. Here are some of the key benefits of patenting:

- **Exclusive rights and protection**
    - **Market exclusivity:** A granted patent provides the inventor with exclusive rights to the invention for a set period (typically 20 years from the filing date). This means no one else can make, use, sell, or distribute the patented invention without permission, allowing the patent holder to control the market for the invention.
    - **Prevention of copying:** Patents prevent competitors from copying or imitating the patented invention, which can help the patent holder maintain a competitive advantage and market share.
- **Commercialization opportunities**
    - **Licensing and royalties:** Patentees can license their inventions to others, generating revenue through royalty agreements. Licensing can be an effective way to monetize an invention without needing to produce or market the product themselves.
    - **Partnerships and collaborations:** A patent can serve as a valuable asset in negotiating partnerships, joint ventures, or collaborations with other businesses, providing opportunities for expanded market reach and additional resources.
    - **Product Differentiation:** Having a patented invention can differentiate a company's products or services in the market, potentially making them more attractive to customers, investors, and partners.
- **Increased business valuation**
    - **Intellectual property asset:** A patent is a form of IP that can be valued as an asset on a company's balance sheet. This can enhance the company's valuation, which is particularly important in mergers, acquisitions, or attracting investors.
    - **Investor confidence:** Having patents can signal to investors that a company is innovative and has valuable, protectable IP. This can increase investor confidence and potentially lead to better funding opportunities.

- **Competitive advantage**
  - **Market monopoly:** Patents provide a temporary monopoly in the market, allowing the patent holder to benefit from their invention without competition for the duration of the patent term. This can lead to higher profit margins and market leadership.
  - **Blocking competitors:** A strong patent portfolio can block competitors from entering the market with similar technologies or products, preventing them from gaining a foothold in the same space.
  - **Technological leadership:** Owning patents can help position a company as a leader in innovation within its industry, fostering brand recognition and customer loyalty.
- **Encouragement of innovation**
  - **Protection of ideas:** Patents encourage inventors to disclose their ideas to the public, knowing they will have legal protection for their inventions. This disclosure benefits society by making the information publicly available, which can spur further innovation and development in the field.
  - **Incentive for R&D:** The ability to patent inventions provides companies with an incentive to invest in research and development (R&D), knowing that they can protect and commercialize their innovations.
- **Defensive use**
  - **Patent portfolio:** A strong patent portfolio can be used defensively to protect against infringement lawsuits. Owning patents can act as a deterrent against competitors who might otherwise infringe on your IP.
  - **Cross-licensing:** Companies with complementary patents may enter into cross-licensing agreements, where each party grants the other the right to use their patents. This can avoid costly litigation and provide mutual benefits.
  - **Freedom to operate:** Patents help ensure that a company can operate without the risk of infringing on others' patents. A "freedom to operate" (FTO) analysis can be conducted to ensure that a company's products or processes do not violate existing patents.
- **Global protection**
  - **International coverage:** Patents provide a mechanism for obtaining protection in multiple countries. While patents are territorial, systems like the PCT allow inventors to file a single international patent application and seek protection in multiple jurisdictions, making it easier to secure global coverage for an invention.
  - **Recognition and credibility:** A granted patent can lend credibility to an inventor or company, enhancing its reputation in the global marketplace and attracting interest from international partners, distributors, and investors.
- **Tax benefits**
  - **Patent box regimes:** Some countries offer tax incentives for businesses that generate income from patented inventions, often

referred to as "patent box" regimes. These tax incentives can reduce the effective tax rate on income derived from patented technologies, providing financial benefits to patent holders.
  - **Tax deductions for R&D:** Patent-related research and development activities may qualify for tax credits or deductions in certain jurisdictions, further incentivizing innovation.
- **Protection of branding and identity**
  - **Trademark synergy:** In some cases, a patented product or technology can be associated with a specific brand, strengthening the company's brand identity. A patented product can be marketed as unique, further distinguishing it from competitors in the eyes of consumers.
- **Legal recourse**
  - **Infringement remedies:** Patent holders have the legal right to take action against any party that infringes on their patent. This can involve seeking damages, injunctions to stop the infringement, or other legal remedies, providing a way to protect the commercial value of the invention.
  - **Deterrent effect:** The mere existence of a patent can deter others from attempting to copy or infringe on the invention, reducing the likelihood of costly legal disputes.

Patenting offers significant benefits by protecting inventions, enabling commercialization, and providing a competitive edge in the market. It fosters innovation, encourages investment in research and development, and provides legal avenues for enforcing IPR. While the process of obtaining and maintaining patents can be costly and complex, the potential rewards, both financial and strategic, often outweigh the challenges, making patents a valuable asset for individuals and businesses alike.

4.7 INTERNATIONAL PATENTING

International patenting refers to the process of seeking patent protection for an invention in multiple countries or regions outside the inventor's home jurisdiction [90]. Since patents are territorial rights, a patent granted in one country does not provide protection in others. International patenting allows inventors and businesses to secure their IPR in multiple countries, providing a broader scope of protection and enabling global commercialization.

Here's an overview of the key aspects of international patenting:

- **Patent cooperation treaty (PCT)**
  - **Unified application process:** The PCT is an international treaty that simplifies the process of filing for patent protection in multiple countries. By filing a single PCT application with the WIPO, inventors can seek patent protection in over 150 contracting states

(countries and regions) without having to file separate applications in each jurisdiction.

  - **Two phases:** The PCT process consists of two phases:
  - **International phase:** The applicant files a single application, which is processed by WIPO. During this phase, the applicant receives an international search report (ISR) and an optional written opinion on patentability, helping them assess the likelihood of obtaining patents in various countries.
  - **National/regional phase:** After the international phase, the applicant must enter the national or regional phase by filing with the patent offices of the countries or regions where they seek protection. The applicant has up to 30 or 31 months from the priority date to enter the national phase.
  - **Advantages:** The PCT system delays the costs and decisions associated with filing in individual countries, giving applicants more time to assess the commercial viability of their invention, secure funding, or assess competition before committing to national filings.
- **European patent convention** (EPC)
  - **European patent:** The EPC allows inventors to file a single patent application with the European Patent Office (EPO) to seek protection in multiple European countries. Once granted, the European patent can be validated in the member states of the EPC, and the patent holder can enforce it in those countries.
  - **Regional approach:** The European patent system streamlines the process of obtaining protection across multiple countries, reducing the complexity of filing separate applications in each jurisdiction. However, after the patent is granted, it must be validated in each country where protection is desired, and maintenance fees must be paid in each country.
  - **Unified patent court** (UPC): The UPC is a court system that will provide a unified forum for patent litigation across participating European Union countries. This aims to make the enforcement of European patents easier and more cost-effective.
- **National filings**
  - **Direct national applications:** In some cases, inventors may choose to file directly with the patent office of the countries in which they seek protection. This is typically done when there is no regional or international filing system available or when protection in specific countries is required.
  - **Cost and complexity:** Filing national applications in multiple countries can be expensive, as it involves separate filings, translations, and compliance with each country's legal and procedural requirements. It also requires paying filing and examination fees in each jurisdiction.

- Regional patent systems
  - **ARIPO (African Regional Intellectual Property Organization):** ARIPO offers a system that allows inventors to file a single patent application and seek protection in multiple African countries. This system simplifies the patent process for inventors looking to protect their inventions across the region.
  - **OAPI (Organisation Africaine de la Propriété Intellectuelle):** OAPI provides a similar system in which inventors can file a single application and gain protection in all member states of OAPI, which includes many French-speaking African countries.
  - **ASEAN patent cooperation:** The Association of Southeast Asian Nations (ASEAN) has discussed the possibility of creating a regional patent system to streamline patent protection in the region, though such a system is not yet fully implemented.
- Costs and considerations
  - **Translation costs:** One of the major costs in international patenting is the translation of the patent application into the official languages of the jurisdictions where protection is sought. For example, a European patent may need to be translated into German, French, or other languages, which can be expensive.
  - **Filing fees:** Filing fees for international patent protection can be significant, especially if the applicant seeks protection in many countries. Fees vary by jurisdiction and include application fees, examination fees, and maintenance fees.
  - **Maintenance fees:** After a patent is granted, maintenance fees must be paid regularly to keep the patent in force. These fees can vary by country and can become quite expensive, especially if protection is sought in multiple countries.
- Challenges of international patenting
  - **Different patent laws:** Patent laws, procedures, and requirements differ from country to country. For example, some countries may have stricter requirements for patentability (such as what constitutes an inventive step or the scope of patentable subject matter), which can affect the success of an application.
  - **Enforcement:** Enforcing patents internationally can be difficult. Patent enforcement is a national issue, and each country has its own legal system for resolving patent disputes. While the PCT and EPC provide some consistency in application, enforcement may still require navigating complex legal systems in different countries.
  - **Patent term and duration:** The duration of patent protection may vary slightly between jurisdictions, and the rules governing supplementary protection certificates (SPCs) or patent extensions differ from country to country. This can impact the commercial life of a product in different markets.

- **Strategic considerations**
  - **Target markets:** When deciding where to file for patent protection, businesses should consider the commercial importance of each market. For example, an inventor might prioritize countries with large markets, such as the United States, China, the European Union, or Japan, where protection is critical for market entry.
  - **Competitor activity:** Filing in jurisdictions where competitors are active or where counterfeiting is a concern may be important for protecting market share. Companies often focus on regions where they are likely to face competition or infringement issues.
  - **Budget and resources:** The cost of filing and maintaining patents in multiple countries can be substantial. Small businesses and startups may need to prioritize certain jurisdictions or regions based on their available budget and commercial goals.
- **Global patenting trends**
  - **China and India:** These countries have become increasingly important in global patenting, both as markets and as innovation hubs. Patent filings in China, in particular, have grown rapidly, and companies looking to enter the Chinese market often seek patent protection there.
  - **Software and AI patents:** The global patent landscape is evolving with new areas of technology, such as software, AI, and biotechnology, raising new challenges in terms of patent eligibility and protection.
  - **Green technologies:** There is a growing focus on patents related to renewable energy and environmental technologies, as governments and businesses invest in solutions to address climate change.

4.8 INTERNATIONAL COOPERATION ON INTELLECTUAL PROPERTY

International cooperation on IP refers to the collaborative efforts between countries, international organizations, and regional bodies to harmonize IP laws, improve enforcement, and facilitate the global protection of IPR [91]. This cooperation helps to ensure that inventors, creators, and businesses can protect their innovations and creations across borders in an increasingly globalized world.

Here are the main frameworks, treaties, and initiatives that promote international cooperation in IP:

- **World Intellectual Property Organization (WIPO)**
  - **Role of WIPO**
    WIPO, a specialized agency of the United Nations, is the central international organization for the promotion and protection of IP

worldwide. It works to harmonize IP laws across different jurisdictions and provides services, legal frameworks, and guidance on IP protection and enforcement.

- **Global IP services**
  WIPO offers several international services for IP protection, including:
  - **Patent cooperation treaty (PCT):** Facilitates international patent filings by allowing inventors to file a single patent application that can be recognized by multiple countries.
  - **Madrid system:** A global system for the registration of trademarks, allowing businesses to file a single application for trademark protection in multiple countries.
  - **Hague system:** Provides a simplified process for the international registration of industrial designs.
  - **Lisbon system:** Covers the international registration of GIs, helping protect products that have a specific geographical origin.

- **World Trade Organization (WTO) and the TRIPS Agreement**
  - **Trade-Related Aspects of Intellectual Property Rights (TRIPS)**
    The TRIPS Agreement, which is part of the WTO, establishes minimum standards for the protection and enforcement of IPR across its member countries. TRIPS requires all WTO members to comply with international IP standards and enforces the protection of patents, copyrights, trademarks, and other IP rights.
  - **Key Provisions of TRIPS:**
    - **Patent Protection:** TRIPS mandates a minimum 20-year protection period for patents.
    - **Copyright Protection:** Copyright protection must extend to literary, artistic, and scientific works.
    - **Enforcement Mechanisms:** The agreement sets out provisions for the enforcement of IP rights, including border measures to prevent counterfeit goods from entering trade.
    - **Flexibilities:** TRIPS allows for certain flexibilities, such as exceptions for public health emergencies (e.g., the Doha Declaration on public health).

- **European Patent Organization (EPO)**
  - **Role of EPO**
    The EPO is responsible for granting European patents under the European Patent Convention (EPC). While it does not grant a "European patent" as a single unit, it provides a streamlined process for inventors to obtain patents that are valid in multiple European countries by filing a single application.
  - **Regional cooperation**
    The EPC allows for greater efficiency in obtaining patent protection in Europe, reducing the need for separate national filings.

The EPO also collaborates with other patent offices globally to harmonize procedures and standards.

- **Patent cooperation treaty (PCT)**
  - **International patent filing**
    The PCT, administered by WIPO, allows inventors to file a single international patent application that can be recognized by over 150 contracting states. The PCT system simplifies the process of seeking patent protection in multiple countries by allowing a single filing, followed by entry into national or regional phases in individual countries or regions.
  - **Global harmonization**
    The PCT promotes international cooperation by providing a standardized procedure for patent filing and examination, helping inventors and businesses streamline their patent applications across multiple jurisdictions.
- **Regional IP cooperation**
  - **African Regional Intellectual Property Organization (ARIPO)**
    ARIPO is a regional organization that allows inventors to file a single patent application and seek protection in multiple African countries. It provides a more cost-effective and efficient way to secure patent protection across the region.
  - **Organisation Africaine de la Propriété Intellectuelle (OAPI)**
    OAPI offers a similar service for French-speaking African countries, allowing inventors to file a single application for protection in all OAPI member states.
  - **ASEAN (Association of Southeast Asian Nations)**
    ASEAN countries are exploring the possibility of a unified regional patent system to simplify the patenting process within Southeast Asia.
- **Geographical indications (GIs) and traditional knowledge**
  - **Geographical indications (GIs)**
    International cooperation has also focused on protecting GIs, which identify products as originating from a specific place, where a given quality, reputation, or characteristic is essentially attributable to that location. WIPO administers the Lisbon System for the international registration of GIs.
  - **Traditional knowledge**
    Efforts to protect traditional knowledge, cultural expressions, and genetic resources have gained traction in international IP law. WIPO's Intergovernmental Committee on Intellectual Property and Genetic Resources, Traditional Knowledge, and Folklore works on developing legal frameworks to protect these assets.
- **Bilateral and multilateral IP agreements**
  - **Free trade agreements (FTAs)**
    Many countries include provisions on IP in their bilateral and multilateral trade agreements. These agreements often go beyond the

TRIPS standards and provide additional protections for IP, such as stronger enforcement mechanisms, longer terms of protection, and stricter patentability criteria.

- **Example**
  The Comprehensive and Progressive Agreement for Trans-Pacific Partnership (CPTPP) includes enhanced IP provisions, such as stronger protections for copyrights, patents, and trademarks, along with commitments to enforce IP rights more effectively.
- **Bilateral IP agreements**
  Countries may enter into bilateral agreements that address specific IP issues, including patent recognition, copyright enforcement, and trademark protection.

- **IP enforcement and anti-counterfeiting cooperation**
  - **Customs and border enforcement**
    International cooperation is vital for preventing the importation and exportation of counterfeit goods. WIPO and national patent offices collaborate with customs authorities to detect and block the trade of counterfeit and pirated goods.
  - **Interpol and national law enforcement**
    Interpol works with national law enforcement agencies to combat IP crimes, including counterfeiting and piracy. International coordination is essential to tackle the global nature of IP infringement.
- **Capacity building and technical assistance**
  - **WIPO's capacity-building programs**
    WIPO provides technical assistance to developing countries to strengthen their IP systems, build human capacity, and improve the enforcement of IP laws. This helps create a more level playing field for countries with limited resources to participate in the global IP system.
  - **IP education and awareness:**
    Many countries, especially in the developing world, benefit from WIPO's training programmes and awareness campaigns, which help to improve understanding of the importance of IP and its role in economic development.
- **Emerging trends in international cooperation**
  - **AI and digital IP**
    As technology evolves, there is increasing international cooperation in the regulation and protection of emerging technologies such as AI, blockchain, and digital media. These areas raise new challenges in IP law, and international cooperation is crucial for developing harmonized standards and legal frameworks.
  - **Environmental and green technologies**
    Global IP cooperation is also focusing on promoting the protection of innovations in renewable energy and environmental technologies. International agreements and initiatives encourage

the sharing of green technologies to address global environmental challenges.

International cooperation on IP is essential for ensuring that creators, inventors, and businesses can protect their innovations across borders. Through organizations like WIPO, regional systems, and international treaties such as TRIPS, countries work together to harmonize IP laws, streamline procedures, and enhance enforcement mechanisms. This cooperation fosters innovation, encourages global trade, and ensures that IP rights are respected worldwide. As new challenges arise in areas such as digital technology and climate change, continued collaboration will be crucial to adapting the global IP system to meet the needs of the modern world.

4.9 PATENTING UNDER PCT

PCT is an international treaty that provides a unified procedure for filing patent applications to seek protection in multiple countries [92]. It simplifies the process of applying for patents in several jurisdictions, allowing inventors to file a single application with a single authority (WIPO) and later extend it to various countries or regions.

4.9.1 Key features of the PCT system

- **Single international application**
  The PCT allows applicants to file a single international patent application that can later be extended to multiple countries or regions. This is more efficient than filing separate applications in each country.
- **International phase and national/regional phase**
  The PCT procedure consists of two main phases:
  - **International phase:** Involves the filing of the international application and the subsequent international search.
  - **National/regional phase:** The applicant enters the national or regional patent offices of individual countries or regions to continue the patent examination process.
- **Timeframe for decision**
  The PCT provides up to **30 or 31 months** from the priority date (the date of the first filing) to decide in which countries or regions the applicant wants to pursue patent protection. This extended time frame allows applicants to assess the commercial viability of their invention before incurring significant costs.

4.9.2 Procedure for filing a patent under the PCT

- Filing the international application
  - **Where to file**
    The international application can be filed with a national patent office (if it is a **receiving office**) or directly with WIPO (through an **International Receiving Office**).
  - **Documents required**
    - A description of the invention.
    - Claims that define the invention.
    - Drawings (if necessary).
    - An abstract summarizing the invention.
    - Priority claim (if applicable) based on an earlier national filing.
  - **Language**
    The application can be filed in one of WIPO's official languages (English, French, or Spanish). If filed in another language, a translation may be required later.
  - **Filing fees**
    The applicant must pay filing fees, including a basic fee and additional fees based on the number of claims and the number of designated countries.
- **International search**
  - **International searching authority (ISA)**: Once the application is filed, an **International searching authority** (ISA) conducts an international search to find relevant prior art (existing patents, publications, etc.) and prepares an **International Search Report (ISR)**.
  - **Written opinion**: The ISA also issues a **written opinion** on the patentability of the invention, which includes an assessment of novelty, inventive step, and industrial applicability.
  - **Search results**: The ISR and Written Opinion are sent to the applicant, who can use this information to decide whether to proceed with the application in various countries.
- **International publication**
  - **Publication**: The international application is published by WIPO 18 months from the priority date (the filing date or earlier date of a first patent application).
  - **Content of publication**: The published document includes the full application, the ISR, and the Written Opinion. The publication is accessible to the public and can be viewed in the WIPO PATENTSCOPE database.
- **Optional supplementary international search**
  - The applicant can request an **optional supplementary international search** to obtain a more thorough prior art search by another authority. This step can provide additional insights into the novelty of the invention before entering the national or regional phase.

- **National/regional phase entry**
  - After the international phase, the applicant must enter the **national or regional phase** to seek patent protection in specific countries or regions. The applicant has up to **30 or 31 months** from the priority date to file the necessary documents and fees with the national or regional patent offices.
  - **Required documents:** Depending on the jurisdiction, the applicant may need to provide translations, amend claims, and pay national filing fees.
  - **Examination:** Each national or regional office will examine the application according to its own patent laws. Some offices may request additional documents or clarifications during the examination process.
- **Grant of patent**
  - If the application meets the patentability criteria (novelty, inventive step, and industrial applicability), the national or regional office grants the patent. The applicant then has exclusive rights to the invention within the granted jurisdiction.
- **Maintenance fees**
  - Once a patent is granted, the applicant must pay **maintenance fees** to keep the patent in force. The fees vary by jurisdiction and must be paid regularly (typically annually) to maintain the patent's validity.

4.9.3 Advantages of the PCT system

- **Simplified process**
  The PCT allows applicants to file a single international application instead of filing separate applications in multiple countries, saving time and reducing administrative complexity.
- **Extended timeframe**
  Applicants have up to 30/31 months to decide which countries or regions they want to seek protection in, providing more time to evaluate the invention's commercial potential and make strategic decisions.
- **International search report (ISR)**
  The ISR and Written Opinion give applicants valuable insights into the potential patentability of their invention before proceeding with the national phase, reducing the risk of spending resources on jurisdictions where the invention may not be patentable.
- **Cost efficiency**
  Although the PCT application incurs initial filing fees, it can be more cost-effective than filing separate national applications upfront, especially for applicants seeking protection in multiple countries.

- **Global reach**
  The PCT provides access to patent protection in over 150 countries, making it an essential tool for inventors and companies seeking to protect their inventions internationally.
- **Post-filing flexibility**
  The applicant can modify the application, amend claims, or even abandon the application in certain jurisdictions without affecting the overall PCT application.

4.9.4 Disadvantages of the PCT system

- **High initial costs**
  While the PCT simplifies the filing process, it can still be costly, especially for applicants seeking protection in multiple countries, as fees for national phase entry and translations can add up.
- **No immediate patent grant**
  The PCT does not result in an international patent grant; applicants must enter the national or regional phase in each jurisdiction where protection is sought. This can still involve a lengthy process of examination and approval.
- **Complexity in the national phase:**
  Although the PCT system simplifies initial filings, the national phase entry process requires navigating the patent laws of multiple jurisdictions, which can be complex and vary significantly.

The **PCT** offers an efficient and flexible process for inventors and businesses seeking international patent protection. By filing a single application, applicants gain access to a broad range of countries and regions and benefit from an international search and valuable insights into the patentability of their invention. While the PCT system can be costly and does not guarantee an immediate grant of patents, it provides a strategic framework for managing international patent filings and ensuring broader protection across global markets.

4.10 SUMMARY

The chapter on **Intellectual Property Rights and the Patenting Process** provides a comprehensive overview of how IP, particularly patents, protects inventions and fosters innovation. It explains the key criteria for patentability: novelty, inventive step, and industrial applicability and outlines the steps involved in the patenting process, from filing an application to the examination and eventual granting of the patent. The chapter also highlights the role of international systems like the **PCT** and regional patenting options in simplifying global patent protection. While patents offer significant benefits, such as exclusive rights and commercial opportunities, the process can be costly, complex, and time-consuming, with challenges in navigating legal requirements and enforcing patents across jurisdictions.

REVIEW QUESTIONS

1. What is the significance of Intellectual Property Rights (IPR) in protecting innovations and creations?
2. Define Intellectual Property (IP) and explain its importance in fostering creativity and innovation.
3. List and briefly describe the different types of Intellectual Property Rights (IPR).
4. What is the difference between **patents** and **trademarks**?
5. Explain the purpose of **copyrights** and the types of works they protect.
6. What are **design rights**, and how do they differ from patents?
7. Describe the concept of **trade secrets** and how they are protected under IPR.
8. What are **Geographical Indications** (GIs), and how do they benefit local producers?
9. What is **Plant Variety Protection**, and why is it important for agricultural innovations?
10. Explain **Integrated Circuit Layout Design Rights** and their role in protecting semiconductor designs.
11. What are the key steps involved in the **patenting process?**
12. Why is it important to **identify the invention** before filing a patent application?
13. What is the role of a **patent search** in the patenting process, and how does it help in identifying prior art?
14. Describe the process of **drafting a patent application**. What are the essential components of a patent application?
15. What happens after the **filing of a patent application?**
16. What is **patent examination**, and how does it determine the patentability of an invention?
17. Explain the importance of **publication of the patent application** in the patenting process.
18. What are the possible outcomes after the **grant of a patent**?
19. What responsibilities does the patent holder have regarding the **maintenance of the patent?**
20. What are some common challenges in **identifying novelty and inventiveness** during the patent process?
21. Why is **patent law complexity** a significant challenge for inventors and patent applicants?

22. How do **high costs** affect the patenting process, and what strategies can inventors use to manage these costs?
23. What are the primary **benefits of patenting** an invention?
24. How does obtaining a patent encourage **innovation** and contribute to economic growth?
25. How does **international patenting** differ from national patenting, and why is it important for global protection?
26. What are the challenges of obtaining patent protection in multiple countries, and how does the **Patent Cooperation Treaty** (**PCT**) help address them?
27. Explain the concept of **international cooperation on Intellectual Property**. How does it facilitate global protection and enforcement of IP rights?
28. What are the key features of the **PCT system**, and how does it streamline the patenting process across multiple countries?
29. Describe the **procedure for filing a patent under the PCT** and its significance in international patent protection.
30. What are the **advantages** and **disadvantages** of the PCT system for patent applicants?

Chapter 5

Patent rights and intellectual property management

5.1 INTRODUCTION

In today's rapidly evolving global economy, Intellectual Property Rights (IPR) play a pivotal role in fostering innovation, protecting creative works, and encouraging economic growth. Among the various forms of IPR, patent rights are essential mechanisms for safeguarding technological advancements and incentivizing inventors to bring new ideas to market. As industries worldwide become increasingly dependent on cutting-edge technologies, the importance of patents in ensuring exclusive rights to new inventions has grown significantly.

This chapter explores the advancements in patent rights and the evolving landscape of intellectual property (IP) management. It delves into the scope of patent rights, examining how these rights protect innovations and the challenges associated with enforcing them. The chapter also highlights the critical role of licensing and technology transfer in facilitating the commercialization of patents and broadening their impact. Additionally, the significance of patent information and databases is discussed, as these tools are vital for conducting patent searches and tracking technological innovations [93].

Furthermore, the chapter covers the importance of Geographical Indications (GIs), which protect the reputation of products linked to specific regions, and the administration of the patent system, with a focus on global cooperation to ensure patents are granted and enforced efficiently. As the world of innovation continues to expand, new developments in IPR, particularly in biological systems, computer software, and traditional knowledge, present both opportunities and challenges for patent law. Through these discussions, the chapter provides a comprehensive understanding of the current trends and future directions in patent rights and IP management.

DOI: 10.1201/9781003485827-5

5.2 PATENT RIGHTS: SCOPE OF PATENT RIGHTS

Patent rights are exclusive legal protections granted to inventors for their inventions, providing them with the right to exclude others from making, using, selling, or distributing the patented invention without permission [94]. These rights are typically granted for a fixed period, usually **20 years** from the filing date, during which the patent holder can commercially exploit their invention or license it to others.

The **scope of patent rights** is defined by the **claims** of the patent, which describe the specific boundaries of the invention's protection. These claims must meet the core requirements for patentability, which include:

- **Novelty:** The invention must be new and not have been disclosed publicly in any form before the filing date. This ensures that patents only protect truly innovative ideas.
- **Non-obviousness (Inventive Step):** The invention must not be obvious to someone skilled in the relevant field, based on prior art. This requirement ensures that patents do not cover trivial modifications or obvious extensions of existing knowledge.
- **Utility (Industrial Applicability):** The invention must be useful and capable of being applied in industry, meaning it must have practical value or application in real-world settings.
- **Full Disclosure:** The inventor must fully disclose how the invention works, allowing others skilled in the field to reproduce it once the patent expires. This disclosure fosters further innovation and knowledge dissemination.

Key aspects of patent rights

- **Exclusive rights:** The patent holder has the exclusive right to make, use, sell, or license the invention. This exclusivity allows inventors to recoup their investment in research and development and profit from their inventions.
- **Territorial nature:** Patent rights are territorial, meaning they are only valid in the countries or regions where the patent is granted. A patent granted in one country does not automatically apply to others, which is why inventors often seek international protection through mechanisms like the **Patent Cooperation Treaty (PCT)**.
- **Patent infringement:** If someone else uses, makes, or sells a patented invention without permission, the patent holder can take legal action for infringement. Patent holders must actively enforce their rights, which may involve litigation or negotiation.
- **Exceptions to patent rights:** There are certain exceptions and limitations to patent rights, including.
- **Research exemption:** In many jurisdictions, patents do not prevent others from using the invention for experimental or research purposes.

- **Compulsory licensing:** In some cases, governments can issue compulsory licenses, allowing other to use a patented invention without the patent holder's consent, often in the interest of public health or national security.
- **Patent expiration and public domain:** Once the patent expires, the invention enters the **public domain**, meaning anyone can use it without seeking permission from the patent holder. This encourages further innovation and allows the invention to benefit society at large.

Global patent protection

Due to the territorial nature of patents, inventors seeking international protection must apply for patents in each country where they wish to secure rights. The **PCT** facilitates this process by allowing a single international application that can later be pursued in individual countries. However, the ultimate decision on granting a patent lies with each national patent office.

The scope of patent rights is defined by the **novelty, non-obviousness,** and **utility** of an invention, with protection granted exclusively to the inventor for a specific period. While these rights provide substantial commercial advantages, they also encourage further technological advancements by ensuring that knowledge is shared after the patent expires.

5.3 LICENSING AND TRANSFER OF TECHNOLOGY

Licensing and **technology transfer** are essential mechanisms that allow the dissemination of innovative technologies and ideas, promoting commercialization and fostering collaboration between different entities [95]. Both concepts are closely tied to patent rights, providing ways for inventors, organizations, and institutions to share their innovations with others while retaining control over IP.

5.3.1 Licensing

Licensing is the process through which a patent holder (the **licensor**) grants permission to another party (the **licensee**) to use, make, sell, or distribute their patented invention under agreed terms and conditions. Licensing enables the patent holder to monetize their IP without having to engage directly in manufacturing or sales.

5.3.1.1 Types of licensing

1. **Exclusive licensing**

   Exclusive licensing is a type of patent licensing agreement in which the patent holder (the **licensor**) grants exclusive rights to a single licensee, meaning the licensee is the only entity authorized to use, make, sell, or distribute the patented invention within a specified territory or market. In this arrangement, the licensor cannot grant the same rights to other parties, nor can they exploit the invention themselves in the same domain during the term of the license.

Key features of non-exclusive licensing

- **Sole rights to use the patent**
  The licensee is granted the exclusive right to exploit the patent, which typically includes the right to make, use, sell, or distribute the patented invention. This means no other company or individual, including the patent holder, can use the patent within the licensed territory or field.
- **Limited geographic or market scope**
  Exclusive licenses can be territorial, applying to specific geographic regions or market segments. For example, the licensor might grant exclusive rights for the patent in the **European Union** but retain the rights in the **United States.**
- **Royalty or lump-sum payments**
  The exclusive licensee often agrees to pay the licensor in the form of **royalties** (a percentage of sales or revenue generated by the invention) or **lump-sum payments** for the exclusive right to use the patent.
- **Licensor's restrictions**
  The licensor typically agrees not to grant further licenses to other parties within the licensed territory, nor will they use the patented invention themselves within that market.
- **Enforcement of patent rights**
  In many cases, the exclusive licensee may also have the right to enforce the patent in case of infringement, though this depends on the terms of the licensing agreement. The licensor may still retain the right to intervene in legal actions.

Advantages of exclusive licensing

- **Monetization without Manufacturing**
  Exclusive licensing allows the patent holder to generate income from their invention without having to invest in manufacturing, marketing, or distributing the product themselves.
- **Incentivizes the Licensee**
  Exclusive licensing motivates the licensee to invest heavily in the development, marketing, and commercialization of the invention, as they have exclusive rights to the patent in the specified territory or market.
- **Higher licensing fees**
  Due to the exclusivity of the arrangement, the licensor can often command higher licensing fees, including upfront payments and royalties, compared to non-exclusive licensing.

- **Focus on strategic partnerships**
  Exclusive licensing can foster closer relationships between the licensor and the licensee, often leading to long-term strategic partnerships for mutual benefit.
- **Maximized use of the invention**
  By granting exclusive rights to a dedicated licensee, the patent holder can ensure that their invention is used to its full commercial potential, benefiting both parties.

Disadvantages of exclusive licensing

- **Limited control**
  The licensor gives up control over the use of the patent within the licensed territory or field, relying on the licensee to manage commercialization effectively.
- **Risk of dependence**
  If the licensee does not succeed in commercializing the invention, the licensor may lose out on potential revenue from other sources or markets.
- **Difficult to revoke**
  Once an exclusive license is granted, it is typically difficult for the licensor to revoke the agreement without legal complications, even if the licensee fails to meet agreed-upon obligations.
- **Potential for market overlap**
  In cases where the licensee's market does not align with the licensor's original vision or strategy, there can be misalignment in the exploitation of the invention, reducing its potential success.
- **Risk of Litigation**
  If the exclusive licensee believes that their rights are being infringed upon, they may take legal action, potentially involving the licensor in litigation even if the licensor did not directly infringe.

When to consider exclusive licensing

Exclusive licensing is often used when:

- The patent holder wants to maximize the commercial potential of their invention through a dedicated partner.
- The patent holder lacks the resources to bring the invention to market themselves but believes the invention has high commercial value.
- The licensee has specialized knowledge, capabilities, or resources that can effectively exploit the patented invention.
- The patent holder wants to maintain long-term relationships with a single entity to drive the success of the invention.

Exclusive licensing offers a powerful tool for patent holders to monetize their inventions while ensuring that the invention is developed and commercialized effectively. While it offers significant benefits, such as higher revenues and a more motivated licensee, it also requires careful consideration of potential risks and limitations. A well-structured exclusive license agreement can foster a mutually beneficial partnership that leads to the successful commercialization of the invention and creates value for both the licensor and licensee.

2. **Non-exclusive Licensing**
   Non-exclusive licensing is a type of patent licensing agreement where the patent holder (the **licensor**) grants the right to use, make, sell, or distribute the patented invention to multiple parties (the **licensees**) simultaneously. Unlike exclusive licensing, the licensor retains the ability to license the same patent to others or use it themselves.

   Key features of non-exclusive licensing

   - **Multiple licensees:** The licensor can grant rights to several licensees at the same time, allowing widespread use of the patented invention.
   - **Retention of rights by licensor:** The licensor retains the right to use the patent and can continue to license it to other entities without restrictions.
   - **Flexibility:** The terms of a non-exclusive license can vary significantly, depending on the agreement, including royalties, geographic scope, and field of use.
   - **Lower costs for licensees:** Since the license is non-exclusive, licensees often pay lower fees compared to exclusive licensing arrangements.

   Advantages of non-exclusive licensing

   - **Broad market penetration:** By licensing the patent to multiple parties, the licensor can achieve greater market coverage and dissemination of the invention.
   - **Revenue diversification:** Licensing to multiple entities creates multiple revenue streams, reducing dependence on a single licensee.
   - **Retained control:** The licensor retains the ability to use the patented invention and license it further, maintaining greater control over the IP.
   - **Fostering competition:** Non-exclusive licensing allows multiple licensees to compete in the market, potentially driving innovation and reducing costs for consumers.
   - **Lower risk for licensees:** Since non-exclusive licenses are often less expensive, licensees can experiment with the technology without significant financial risk.

Disadvantages of non-exclusive licensing

- **Limited exclusivity for licensees:** Licensees may be less motivated to invest heavily in the development and marketing of the invention, as competitors can also use the same technology.
- **Potential market saturation:** If too many licenses are granted, it could lead to market saturation, reducing the overall value of the invention.
- **Reduced revenue per license:** The licensor may earn lower royalties or fees from individual licensees compared to exclusive licensing agreements.
- **Complexity in management:** Managing multiple license agreements and ensuring compliance can be administratively challenging for the licensor.
- **Risk of dilution:** Widespread use of the invention by multiple licensees may dilute the perceived uniqueness or value of the patented technology.

When to consider non-exclusive licensing

Non-exclusive licensing is suitable when:

- The patent holder wants to maximize market penetration and widespread adoption of the technology.
- The invention has broad applications across different industries or markets.
- The licensor seeks to diversify revenue streams and reduce dependence on a single licensee.
- The invention's value increases with greater usage or network effects, such as software platforms or standard technologies.

Examples of non-exclusive licensing

- **Standard-essential patents (SEPs):** In industries such as telecommunications, non-exclusive licensing is common for technologies that are essential to industry standards, such as 4G or 5G technologies.
- **Open licensing models:** Some licensors offer non-exclusive licenses for their patents to encourage widespread adoption and development, often seen in software and biotech industries.

Non-exclusive licensing is a versatile and effective strategy for patent holders who aim to maximize the reach and utilization of their inventions. By granting rights to multiple licensees, licensors can achieve broader market coverage and generate diversified revenue streams. However, careful management is required to avoid market saturation and ensure that the value of the patented technology is preserved. Non-exclusive licensing fosters collaboration, competition, and innovation, making it a popular choice for many industries.

3. Sole licensing
   Sole licensing is a type of patent licensing arrangement in which the patent holder (the **licensor**) grants exclusive rights to a single licensee to use, make, sell, or distribute the patented invention within a specified territory or field. However, unlike **exclusive licensing**, the licensor retains the right to use the patented invention themselves, but they cannot license it to any other third party during the term of the sole license.

   Key features of sole licensing

   - **Single licensee:** The licensor grants rights to only one licensee, ensuring that no other party (besides the licensor) can use the patent within the licensed scope.
   - **Licensor's retained rights:** The licensor maintains the right to use the patented invention themselves but cannot issue additional licenses to other entities.
   - **Defined scope:** The sole license agreement typically defines specific geographic regions, markets, or fields of use where the licensee can exercise their rights.
   - **Exclusive commercial opportunity for licensee:** The licensee has the sole opportunity to commercialize the invention within the agreed parameters, minimizing direct competition.

   Advantages of sole licensing

   - **Controlled commercialization:** The licensor can partner closely with a single licensee while retaining the ability to use the invention for their own purposes.
   - **Incentivizes the licensee:** Since the licensee does not face competition from other licensees, they are more likely to invest resources in the development and commercialization of the invention.
   - **Licensor's continued use:** The licensor retains the ability to use the invention themselves, which can be advantageous for their own operations or research.
   - **Simpler management:** Managing a sole license is often simpler than managing multiple non-exclusive licenses, as there is only one licensee involved.
   - **Potential for strategic partnerships:** Sole licensing often fosters strong collaboration between the licensor and the licensee, creating opportunities for long-term partnerships.

   Disadvantages of sole licensing

   - **Limited revenue potential:** The licensor forgoes the opportunity to generate income from additional licensees, which could limit the overall financial returns.

- **Dependence on a single licensee:** The licensor's success in monetizing the patent may become overly reliant on the licensee's ability to commercialize the invention effectively.
- **Potential for underutilization:** If the sole licensee fails to maximize the invention's commercial potential, the patent's value may not be fully realized.
- **Restricted licensing opportunities:** During the term of the sole license, the licensor cannot grant licenses to other parties, which could be a drawback in dynamic markets.
- **Complex negotiations:** Establishing terms that satisfy both the licensor and the sole licensee can require complex negotiations, especially regarding the licensor's retained rights.

When to consider sole licensing

Sole licensing is often used when:

- The licensor wants to maintain some control over the use of their patent while granting commercialization rights to a single entity.
- The invention is highly specialized or targeted at a specific market, making a single licensee sufficient for its commercial success.
- The licensor values a strong, strategic partnership with the licensee over broader market penetration.

Examples of sole licensing

- **Pharmaceuticals:** A pharmaceutical company may grant a sole license to a regional partner to manufacture and distribute a patented drug in a specific country while retaining the right to market it in other regions.
- **Academic innovations:** Universities or research institutions may grant sole licenses to companies for commercializing patented technologies while continuing to use the technology for further research.

Sole licensing strikes a balance between exclusive and non-exclusive licensing, offering a single licensee the opportunity to commercialize a patented invention while allowing the licensor to retain usage rights. It is particularly useful when the licensor seeks a focused partnership and wants to ensure that the invention is used effectively without entirely relinquishing control. While it offers significant benefits, such as streamlined management and strategic collaboration, sole licensing also requires careful negotiation to align the interests of both parties and maximize the invention's potential.

5.3.1.2 Benefits of licensing

1. **Revenue generation**
   Licensing IP, such as patents, trademarks, or copyrights, is a strategic method for generating revenue. By granting rights to other entities to

use, manufacture, or sell their IP, licensors can monetize their assets without directly engaging in production or distribution. Licensing provides several avenues for revenue generation, which can be tailored to the specific needs and agreements of the parties involved.

Key revenue streams in licensing

- **Royalty payments**
  - Licensees typically pay the licensor a percentage of sales or profits derived from the licensed IP.
  - Royalties can be structured as a fixed percentage (e.g., 5–10% of sales) or tiered, increasing with higher sales volumes.
- **Upfront payments**
  - Many licensing agreements include an initial lump-sum payment made by the licensee to secure the licensing rights.
  - This provides immediate revenue for the licensor, offsetting the initial costs of developing the IP.
- **Milestone payments**
  - For industries such as pharmaceuticals or technology, license agreements may include payments tied to specific achievements, such as regulatory approvals, market launches, or sales milestones.
  - These payments incentivize both parties to collaborate effectively.
- **Annual fees**
  - Some licensing agreements include fixed annual fees, ensuring steady revenue regardless of sales performance.
  - This approach is often used for software licensing or standard-essential patents.
- **Sub-licensing revenue**
  - In cases where the licensee is allowed to grant sub-licenses, the licensor may receive a portion of the revenue generated from these sub-licenses.
- **Territorial licensing**
  - Licensors can generate revenue by licensing the same IP to different entities in distinct geographic regions, maximizing market reach and income.

Advantages of revenue generation through licensing

- **Low capital investment**
  - Licensing eliminates the need for the licensor to invest heavily in manufacturing, marketing, or distribution. Instead, the licensee assumes these responsibilities.
- **Risk mitigation**
  - By transferring commercialization risks to the licensee, the licensor avoids the financial and operational challenges associated with bringing a product to market.

- **Scalability**
  - Licensing agreements can be structured with multiple licensees across various markets, creating diverse and scalable revenue streams.
- **Recurring income**
  - Royalties and annual fees provide ongoing revenue over the term of the licensing agreement, ensuring a steady cash flow.
- **Global market access**
  - Licensing to international partners enables licensors to tap into global markets, increasing revenue potential without establishing a physical presence.
- **Value maximization**
  - Licensing allows the licensor to fully exploit the commercial potential of their IP across different industries or applications.

Licensing is a powerful tool for generating revenue from IP, offering flexibility, scalability, and low-risk opportunities for monetization. Through royalties, upfront payments, and other revenue streams, licensors can maximize the value of their innovations while focusing on core competencies or new developments. By structuring agreements strategically, licensing can become a cornerstone of a sustainable and profitable IP management strategy.

2. **Market expansion**
   Licensing IP facilitates **market expansion** by enabling the licensor to reach new geographic regions, customer bases, and industry sectors through partnerships with licensees. This strategy allows licensors to extend their market presence without directly investing in infrastructure, marketing, or distribution networks.

   How licensing drives market expansion

   - **Access to new geographies**
     - Licensing agreements often grant licensees the right to use or distribute the IP in specific regions.
     - This is particularly beneficial for entering foreign markets where the licensor may lack a local presence or expertise.
   - **Leveraging local expertise**
     - Licensees typically have established networks, market knowledge, and regulatory familiarity in their regions, allowing the licensor's product or technology to gain traction more quickly.
   - **Broader customer reach**
     - Licensing enables products or services to be marketed to diverse customer bases across different demographics or industries.
   - **Rapid scalability**
     - Licensing accelerates market penetration by allowing multiple licensees to commercialize the IP simultaneously in different territories or sectors.

- **Diversified applications**
    - Inventions or technologies often have multiple applications across various industries. Licensing to specialized firms in these sectors can unlock untapped markets.
- **Compliance with regional regulations**
    - Local licensees are better equipped to navigate regional legal and regulatory requirements, facilitating smoother entry into restricted or highly regulated markets.

Advantages of market expansion through licensing

- **Low-cost market entry**
    - The licensor avoids the costs of establishing a physical presence, manufacturing facilities, or distribution channels in new markets.
- **Risk sharing**
    - Licensees bear the operational and financial risks of entering new markets, reducing the burden on the licensor.
- **Brand recognition growth**
    - Licensing agreements can enhance brand visibility and reputation in new markets, creating opportunities for further expansion or product diversification.
- **Increased revenue streams**
    - Expanding into multiple markets generates additional income through royalties, upfront fees, and milestone payments from licensees.
- **Enhanced innovation**
    - Market feedback from licensees operating in diverse regions or industries can inspire improvements and innovations for the licensed technology or product.

Examples of market expansion through licensing

- **Pharmaceutical industry:** Licensing a patented drug to regional partners for manufacturing and distribution enables pharmaceutical companies to reach global markets.
- **Technology licensing:** Software firms often license their products to local companies in emerging markets, allowing faster adoption.
- **Consumer goods:** Brands license trademarks to international partners to manufacture and sell their products locally, leveraging the partner's market expertise.

Market expansion through licensing is a strategic way for licensors to grow their reach without the complexities and costs of direct market entry. By partnering with licensees who understand local markets and regulations, licensors can effectively introduce their IP to new audiences, creating opportunities for sustained growth and revenue diversification.

3. **Risk mitigation**
Licensing IP provides an effective strategy for mitigating various risks associated with commercialization, market entry, and operational challenges. By transferring certain responsibilities to licensees, licensors can focus on innovation and strategic growth while minimizing financial, operational, and legal risks.

How licensing mitigates risks

- **Reduced financial burden**
  - Licensing eliminates the need for the licensor to invest heavily in manufacturing, marketing, or distribution, as these responsibilities are typically borne by the licensee.
  - This reduces upfront costs and financial exposure, especially for small or resource-constrained entities.
- **Shared commercialization risk**
  - The licensee assumes the risks of bringing the product or technology to market, including development, production, and sales efforts.
  - This allows the licensor to benefit from market success without directly facing the challenges of commercialization.
- **Market entry risk reduction**
  - Entering new geographic regions or industries often involves significant risks, such as regulatory hurdles, cultural differences, and unfamiliar markets.
  - Licensing to local partners mitigates these risks, as licensees typically have the necessary expertise and infrastructure.
- **Avoidance of operational complexity**
  - Managing operations like production, distribution, and marketing in diverse markets can be challenging. Licensing transfers these operational risks to the licensee.
- **Legal and regulatory compliance**
  - Licensees are often better equipped to navigate local laws and regulations, reducing the risk of non-compliance for the licensor.
- **IP protection risks**
  - Licensing agreements can include provisions to ensure that the licensee protects the IP from infringement or misuse, reducing the licensor's enforcement burden.
- **Diversified revenue streams**
  - Licensing to multiple entities in different markets diversifies income sources, mitigating the risk of financial dependency on a single market or product.

Advantages of risk mitigation through licensing

- **Focus on core competencies**
  - By offloading operational risks, licensors can concentrate on innovation, research, and development rather than market operations.
- **Scalability with lower risk**
  - Licensing enables licensors to expand their market reach without the risks associated with direct investment in infrastructure or staffing.
- **Reduced exposure to market volatility**
  - Licensing agreements provide stable revenue streams (e.g., royalties), insulating licensors from fluctuations in market demand or economic conditions.
- **Faster adaptation to market changes**
  - Licensees, being closer to the market, can adapt more quickly to changing conditions, reducing the risk of lagging behind competitors.
- **Protection against ip misuse**
  - Well-drafted licensing agreements include clauses for monitoring and enforcement, ensuring that the licensee uses the IP responsibly and legally.

Examples of risk mitigation in licensing

- **Technology sector:** Tech companies often license their patents to manufacturing partners, avoiding the risks of setting up production facilities.
- **Pharmaceuticals:** Drug developers license their patents to established pharmaceutical firms for manufacturing and distribution, mitigating regulatory and operational risks.
- **Consumer goods:** Brands license their trademarks to regional partners who manage production and distribution in local markets, reducing logistical challenges.

Licensing serves as a powerful tool for mitigating risks associated with the commercialization and expansion of IP. By leveraging the expertise and resources of licensees, licensors can reduce financial exposure, operational complexities, and legal challenges. This collaborative approach ensures that risks are distributed effectively while maximizing the potential benefits of the IP.

4. **Fostering innovation**
   Licensing IP fosters innovation by enabling the sharing, development, and commercialization of new ideas and technologies. Through

licensing agreements, inventors, businesses, and researchers can collaborate, pool resources, and build on existing innovations, accelerating the pace of technological and market advancements.

How licensing fosters innovation

- **Access to advanced technology**
  - Licensing provides licensees with access to cutting-edge technologies or IP, which they can adapt or enhance to create innovative products or solutions.
- **Encouraging research and development** (R&D):
  - Licensing generates revenue for licensors, which can be reinvested into further R&D, fueling a continuous cycle of innovation.
  - Licensees, too, are incentivized to innovate by incorporating licensed IP into their products to gain a competitive edge.
- **Collaboration and knowledge sharing**
  - Licensing agreements often foster partnerships between licensors and licensees, encouraging the exchange of technical expertise and market insights.
  - This collaborative environment promotes innovation across industries and geographies.
- **Overcoming resource limitations**
  - Small or resource-constrained entities can license their IP to larger companies with the capabilities to develop and commercialize it, ensuring that valuable innovations reach the market.
- **Promoting industry standards**
  - Licensing essential patents to multiple companies can establish industry standards, enabling interoperability and encouraging innovation within a unified framework.
- **Reducing redundancy**
  - By licensing existing technologies, organizations can avoid duplicating efforts in R&D, focusing instead on building complementary innovations.
- **Global Reach and Customization**
  - Licensing to regional partners allows innovations to be tailored to local needs and preferences, fostering diverse applications of the same technology.

Advantages of fostering innovation through licensing

- **Accelerated development cycles**
  - Licensees can build on proven technologies, reducing time-to-market for innovative products.
- **Diversified applications**
  - Licensing enables innovations to be adapted for various industries or applications, maximizing their impact and utility.

- **Enhanced competition**
  - Licensing levels the playing field by allowing smaller firms to access advanced technologies, driving competition and innovation in the market.
- **Increased value of IP**
  - As innovations evolve and gain market traction through licensing, the value of the underlying IP increases.
- **Sustainable innovation ecosystem**
  - Licensing supports an ecosystem where inventors, businesses, and researchers can continuously innovate and commercialize ideas collaboratively.

Examples of licensing fostering innovation

- **Technology sector:** Companies licensing software development kits (SDKs) encourage third-party developers to create new applications, enhancing the core technology.
- **Pharmaceutical industry:** Licensing patented drugs to multiple manufacturers leads to the development of improved formulations or alternative uses.
- **Automotive industry:** Licensing battery or autonomous vehicle technology promotes advancements in electric vehicles and smart mobility solutions.

Licensing is a catalyst for fostering innovation, enabling the dissemination and enhancement of IP across industries and markets. By bridging the gap between invention and commercialization, licensing promotes collaboration, accelerates technological progress, and ensures that innovative solutions reach their full potential. This approach not only benefits individual licensors and licensees but also contributes to the overall advancement of industries and societies.

5.3.2 Technology transfer

Technology transfer refers to the process of transferring scientific findings, knowledge, or technologies from one entity to another, typically from research institutions (like universities or R&D labs) to the private sector for commercialization. It plays a crucial role in transforming innovations into practical, market-ready products or services.

Key forms of technology transfer

- **Licensing:** As discussed, licensing is one of the most common forms of technology transfer, where the patent holder grants rights to use their technology.
- **Joint ventures:** In some cases, technology transfer occurs through partnerships or joint ventures, where two or more entities collaborate to develop and commercialize new technologies.

- **Research collaboration agreements:** Universities, research institutions, and companies may enter into agreements to jointly develop new technologies, with the understanding that the resulting IP will be shared or licensed.
- **Spin-offs and startups:** In some cases, research institutions or universities create spin-off companies to commercialize innovations. This type of technology transfer involves the creation of a new business to bring a specific technology to market.

Benefits of technology transfer

- **Accelerated innovation:** By transferring technology from research settings to commercial entities, new products can be developed faster, benefiting society and the economy.
- **Public-private collaboration:** Technology transfer often involves collaboration between public research institutions and private companies, enhancing the commercialization of publicly funded research.
- **Economic growth:** Successful technology transfer contributes to economic growth by creating jobs, fostering new industries, and increasing global competitiveness.
- **Global access to innovation:** Technology transfer facilitates the global spread of new technologies, helping to address global challenges such as healthcare, climate change, and sustainable development.

5.4 PATENT LICENSING IN TECHNOLOGY TRANSFER

In many cases, technology transfer relies on **patent licensing** as the legal mechanism to transfer the rights to use a patented invention. By licensing patents, research institutions, universities, and businesses can ensure that their innovations are used in the commercial sector, which can lead to the development of new products, services, and markets [96].

For instance, universities often patent the results of their research and license these patents to companies that have the resources to bring the inventions to market. This model not only benefits the universities through royalty income but also accelerates the commercialization of innovations that might otherwise remain confined to academic settings.

Patent licensing plays a pivotal role in the process of technology transfer, enabling the commercialization and dissemination of innovative technologies. Through licensing agreements, patent holders (licensors) grant permission to other entities (licensees) to use, develop, manufacture, or sell a patented technology, often in exchange for royalties or other forms of compensation. This process facilitates the movement of technology from creators to users, bridging the gap between research and market application.

Key aspects of patent licensing in technology transfer

- **Enabling commercialization**
  - Licensing allows organizations or individuals to commercialize technologies developed by others, ensuring that innovations reach the market and benefit society.
- **Access to advanced technologies**
  - Patent licensing provides licensees with access to cutting-edge innovations, reducing the need for duplicative R&D efforts and accelerating product development.
- **Revenue generation for inventors**
  - Inventors and research institutions can monetize their patents through licensing agreements, creating a sustainable revenue stream to fund further research.
- **Facilitating collaboration**
  - Licensing agreements often involve partnerships between licensors and licensees, promoting knowledge exchange and joint development efforts.
- **Global dissemination of technology**
  - Licensing patents to entities in different geographic regions ensures that technologies are widely adopted and adapted to meet diverse market needs.

Types of patent licensing in technology transfer

- **Exclusive licensing**
  - The licensee gains exclusive rights to use the patented technology, often within a specific market or region, while the licensor retains ownership of the patent.
- **Non-exclusive licensing**
  - The licensor grants rights to multiple licensees, allowing broader dissemination and adoption of the technology.
- **Sole licensing**
  - Combines elements of exclusive and non-exclusive licensing, where only one licensee is permitted to use the technology, but the licensor may also use it.

Benefits of patent licensing in technology transfer

- **Accelerated innovation**
  - Licensing promotes the adoption and adaptation of new technologies, encouraging innovation across industries.
- **Risk sharing**
  - Licensees assume the risks of commercialization, reducing the financial and operational burden on licensors.

- **Market expansion**
  - Licensing agreements enable technologies to penetrate new markets, including those in developing regions, fostering global growth.
- **Increased return on investment**
  - Licensing ensures that research and development investments yield financial returns, incentivizing further innovation.
- **Encouraging collaboration**
  - Technology transfer through licensing fosters partnerships between academia, industry, and governments, driving economic and technological development.

Challenges in patent licensing for technology transfer

- **Valuation of technology**
  - Determining the appropriate value of a patent can be complex and may lead to disagreements between licensors and licensees.
- **Negotiation complexity**
  - Drafting and agreeing upon licensing terms requires careful negotiation to balance the interests of both parties.
- **Enforcement of agreements**
  - Ensuring compliance with licensing terms and addressing potential infringements can be legally and financially demanding.
- **Cultural and legal barriers**
  - Licensing across borders may encounter challenges related to differing legal systems, cultural practices, and market dynamics.

Examples of patent licensing in technology transfer

- **University research commercialization:** Universities license patents for innovative technologies developed in their labs to private companies for production and market distribution.
- **Technology companies:** Firms like IBM and Qualcomm license their patents for software, hardware, and telecommunications to other companies, fostering industry growth.
- **Pharmaceuticals:** Patent licensing enables the transfer of drug formulations and medical technologies to manufacturers for large-scale production.

Patent licensing is a cornerstone of effective technology transfer, driving the commercialization of innovations and fostering global technological progress. By enabling collaboration, reducing commercialization risks, and ensuring widespread access to advanced technologies, licensing transforms patented inventions into impactful solutions for industries and societies worldwide.

5.5 CHALLENGES IN LICENSING AND TECHNOLOGY TRANSFER

Licensing and technology transfer are critical processes for translating IP into practical applications. However, these processes are often fraught with challenges that can hinder the successful commercialization and dissemination of innovations [97]. Understanding and addressing these challenges is essential for maximizing the benefits of licensing and technology transfer.

Valuation of intellectual property

- **Challenge:** Accurately determining the value of a patent or technology is complex and often subjective.
- **Impact:** Misvaluation can lead to disputes, unbalanced agreements, or missed opportunities.
- **Solution:** Employing professional valuation methods and third-party assessments to ensure fair and transparent pricing.

Negotiation difficulties

- **Challenge:** Reaching a consensus on licensing terms, including royalties, scope, and exclusivity, can be contentious.
- **Impact:** Prolonged negotiations delay the commercialization process.
- **Solution:** Engaging skilled negotiators and mediators to facilitate fair agreements.

Legal and regulatory barriers

- **Challenge:** Licensing agreements often span multiple jurisdictions, each with distinct IP laws and regulations.
- **Impact:** Misalignment with local laws can lead to contract disputes or invalid agreements.
- **Solution:** Ensuring legal compliance through expertise in international IP law and jurisdiction-specific regulations.

Cultural and organizational misalignment

- **Challenge:** Differences in business practices, cultural norms, and organizational goals between licensors and licensees.
- **Impact:** Miscommunication and conflicting expectations can undermine partnerships.
- **Solution:** Building mutual understanding and aligning objectives through clear communication and structured agreements.

Protection of intellectual property

- **Challenge:** Ensuring that licensees do not misuse, infringe upon, or fail to adequately protect the licensed IP.
- **Impact:** Loss of competitive advantage and potential legal battles.
- **Solution:** Incorporating strict IP protection clauses and monitoring mechanisms into licensing agreements.

Risk of dependency

- **Challenge:** Licensors may become overly reliant on a single licensee or region for revenue.
- **Impact:** Financial vulnerability if the licensee underperforms or breaches the agreement.
- **Solution:** Diversifying licensees and regions to spread risks.

Difficulty in technology adaptation

- **Challenge:** Transferring complex or cutting-edge technologies may require significant adaptation or additional development.
- **Impact:** Increased costs and delays in market entry.
- **Solution:** Providing technical support and collaborative development frameworks to assist licensees.

Lack of expertise and resources

- **Challenge:** Smaller entities may lack the expertise or resources to effectively license or transfer technology.
- **Impact:** Limited reach and underutilization of valuable IP.
- **Solution:** Partnering with experienced intermediaries or leveraging technology transfer offices (TTOs).

Market uncertainty

- **Challenge:** Rapidly changing market dynamics and technological obsolescence can reduce the relevance of licensed IP.
- **Impact:** Reduced profitability and short-lived agreements.
- **Solution:** Incorporating flexible terms and ongoing market assessments to adapt to changes.

Enforcement challenges

- **Challenge:** Monitoring compliance with licensing terms and addressing breaches can be resource-intensive.
- **Impact:** High legal costs and potential loss of IP control.
- **Solution:** Establishing clear enforcement mechanisms and dispute resolution processes in the agreement.

Limited technology transfer capacity

- **Challenge:** Organizations in developing regions may lack the infrastructure or expertise to adopt advanced technologies.
- **Impact:** Inefficient utilization of licensed IP.
- **Solution:** Providing training, resources, and capacity-building initiatives to ensure successful adoption.

While licensing and technology transfer present significant opportunities for innovation and market growth, they also involve multifaceted challenges. Addressing these hurdles requires a combination of strategic planning, legal expertise, effective communication, and ongoing collaboration between licensors and licensees. By overcoming these obstacles, stakeholders can unlock the full potential of IP and drive technological and economic progress.

5.6 ROLE OF TECHNOLOGY TRANSFER OFFICES (TTOs)

TTOs play a critical role in bridging the gap between research and commercialization. These specialized entities, typically housed within universities, research institutions, or government agencies, facilitate the transfer of knowledge, innovations, and IP from creators to industries or other stakeholders [98]. Their mission is to maximize the societal and economic impact of research while ensuring fair returns for inventors and institutions.

Key functions of TTOs

- **Identification and evaluation of innovations**
  - **Role:** TTOs identify potentially marketable research outcomes and evaluate their commercial viability.
  - **Activities**
    - Conducting market assessments to determine demand.
    - Assessing the novelty and patentability of innovations.
    - Aligning innovations with industry needs.
- **Protection of intellectual property**
  - **Role:** TTOs safeguard the IPRs of inventors and institutions.
  - **Activities**
    - Filing patents, trademarks, and copyrights.
    - Managing legal documentation and IP portfolios.
    - Ensuring compliance with national and international IP laws.
- **Facilitating licensing agreements**
  - **Role:** TTOs act as intermediaries to negotiate and formalize licensing agreements between inventors and commercial entities.

- **Activities**
  - Drafting and negotiating licensing terms.
  - Structuring agreements, including royalties and revenue sharing.
  - Monitoring compliance with licensing contracts.
- **Promoting technology commercialization**
  - **Role:** TTOs connect researchers with industry partners to bring innovations to market.
  - **Activities**
    - Organizing technology showcases and networking events.
    - Partnering with venture capitalists and startups.
    - Supporting prototype development and proof-of-concept testing.
- **Capacity building and training**
  - **Role:** TTOs educate researchers and stakeholders on IP management and commercialization processes.
  - **Activities**
    - Conducting workshops on patent filing and IP rights.
    - Providing mentorship for startups and spin-offs.
    - Offering guidance on industry collaboration.
- **Revenue generation**
  - **Role:** TTOs ensure that institutions and inventors receive financial returns from commercialized technologies.
  - **Activities**
    - Negotiating equitable revenue-sharing agreements.
    - Managing income from licensing fees and royalties.
    - Reinvesting earnings into research and development.
- **Supporting start-ups and spin-offs**
  - **Role:** TTOs foster entrepreneurial ventures based on institutional research.
  - **Activities**
    - Assisting in business plan development.
    - Providing access to incubation facilities.
    - Securing funding and industry partnerships.
- **Facilitating international collaborations**
  - **Role:** TTOs help researchers and institutions collaborate with global partners for technology transfer.
  - **Activities**
    - Navigating international IP regulations.
    - Building cross-border partnerships.
    - Facilitating technology licensing in international markets.

Importance of TTOs

- **Accelerating innovation**
  - TTOs ensure that groundbreaking research translates into practical applications, fostering innovation across industries.
- **Economic impact**
  - By enabling the commercialization of technologies, TTOs contribute to job creation, economic growth, and societal benefits.
- **Enhancing institutional reputation**
  - Successful technology transfer elevates the status of universities and research institutions, attracting talent and funding.
- **Fostering collaboration**
  - TTOs act as a bridge between academia and industry, facilitating partnerships that benefit both sectors.

Challenges faced by TTOs

- **Resource constraints**
  - Limited funding and staffing can hinder TTO operations.
- **Market uncertainty**
  - Assessing the commercial potential of early-stage technologies can be challenging.
- **Complex negotiations**
  - Licensing agreements require balancing the interests of multiple stakeholders.
- **Awareness and engagement**
  - Encouraging researchers to engage with TTOs and prioritize commercialization can be difficult.

5.7 PATENT INFORMATION AND DATABASES

Patent information and databases are essential tools for managing, searching, and analyzing IP. They serve as repositories of technical, legal, and commercial data related to patents, offering valuable insights for researchers, inventors, businesses, and policymakers [99]. These resources not only facilitate the innovation process but also help avoid duplication of effort and ensure compliance with IP laws.

Components of patent information

- **Technical details**
  - Description of the invention, including drawings, specifications, and claims.

    - Enables understanding of the scope and functionality of the patented technology.
- **Legal information**
    - Details about patent ownership, filing dates, jurisdictions, and legal status.
    - Useful for determining enforceability and territorial rights.
- **Commercial data**
    - Licensing agreements, market applications, and potential competitors.
    - Helps in assessing the market potential and strategic positioning of a patent.

Uses of patent information

- **Prior art searches**
    - Identifying existing patents to determine novelty and avoid infringement.
- **Competitive analysis**
    - Monitoring competitors' patent portfolios to identify trends and strategies.
- **Technology development**
    - Gaining insights into emerging technologies and innovation gaps.
- **Legal compliance**
    - Ensuring that new products or processes do not infringe on existing patents.
- **Market and licensing opportunities**
    - Identifying patents available for licensing or acquisition.

Key patent databases

- **Public databases**
    - **Espacenet** (European Patent Office): Comprehensive global patent data.
    - **USPTO patent database** (United States Patent and Trademark Office): US-specific patent information.
    - **WIPO PATENTSCOPE** (World Intellectual Property Organization): International PCT applications.
- **Commercial databases**
    - **Derwent innovation:** Enhanced patent analysis and visualization tools.
    - **Orbit intelligence:** Comprehensive data for patent analytics and portfolio management.
    - **Pat base:** Multi-lingual, full-text patent search platform.

Benefits of patent databases

- **Accessibility**
  - Provide centralized access to global patent data.
- **Efficiency**
  - Enable quick searches and comparisons of patents.
- **Cost-effectiveness**
  - Public databases offer free access, reducing research costs.
- **Enhanced decision-making**
  - Support R&D, legal, and business strategies with accurate data.

Challenges in using patent databases

- **Data overload**
  - The sheer volume of patent information can be overwhelming.
- **Search complexity**
  - Requires expertise to conduct effective keyword and classification-based searches.
- **Language barriers**
  - Patents filed in multiple languages may pose translation challenges.
- **Incomplete data**
  - Some databases may not have the latest updates or full-text availability.

Role in innovation and research

Patent information and databases empower innovators to build upon existing knowledge, identify new opportunities, and avoid duplicating efforts. They are invaluable for fostering collaboration, advancing technology, and driving economic growth through informed decision-making in IP management.

5.8 NEW DEVELOPMENTS IN IPR

The landscape of IPR is constantly evolving due to technological advancements, globalization, and changes in legal frameworks [100]. Here are some notable developments shaping the future of IPR:

Artificial intelligence (AI) and IPR

- **AI-generated innovations**
  Legal systems are grappling with whether AI-generated inventions can be patented or copyrighted and how to attribute ownership.
- **Example:** Patents for AI-designed molecules in drug discovery.
- **Impact:** Potential updates in IP laws to include AI as co-creators.

Blockchain in IP management

- **Decentralized tracking**
  Blockchain technology is being utilized to create transparent, tamper-proof records of IP ownership and transactions.
- **Example:** Smart contracts for royalty payments in music licensing.
- **Impact:** Improved IP enforcement and reduction in infringement cases.

Expanding scope of geographical indications (GIs)

- **New products and territories**
  More countries are recognizing GIs for non-traditional products like handicrafts, traditional foods, and beverages.
- **Example:** Darjeeling Tea (India) and Champagne (France).
- **Impact:** Increased global trade and protection for local economies.

Digital content and copyright

- **Rise of streaming and NFTs**
  The explosion of digital content and Non-Fungible Tokens (NFTs) has introduced challenges in copyright protection and royalty distribution.
- **Example:** NFT marketplaces for artwork and digital collectables.
- **Impact:** Need for robust copyright policies tailored to digital assets.

Green technology and IPR

- **Eco-innovations**
  Patents for sustainable technologies, such as renewable energy solutions, are receiving special incentives like fast-track examinations.
- **Example:** Patents for solar panels with enhanced efficiency.
- **Impact:** Encouragement of eco-friendly innovations through IPR systems.

Harmonization of international IP laws

- **Global cooperation**
  Efforts like the PCT and agreements through WIPO aim to streamline international filings and enforcement.
- **Example:** Unified Patent Court in the European Union.
- **Impact:** Simplified cross-border IP protection for innovators.

Challenges in biotech and genetic patenting

- **Gene editing and CRISPR**
  Ethical and legal debates around patenting genetic materials and biotechnologies continue to grow.
- **Example:** Patents for CRISPR-Cas9 technology.

- **Impact:** Calls for balance between innovation incentives and public health.

Focus on SMEs and start-ups

- **Incentives and support**
  Governments are creating programmes to help small businesses and startups navigate IPR systems.
- **Example:** Reduced fees for SMEs filing patents or trademarks.
- **Impact:** Increased innovation from smaller entities.

These developments underscore the dynamic nature of IPR and its crucial role in fostering innovation and economic growth. As technology and markets evolve, so must the frameworks governing IP, ensuring fairness, accessibility, and relevance in a rapidly changing world.

5.9 ADMINISTRATION OF THE PATENT SYSTEM

The administration of the patent system is a critical function that ensures the effective protection of IPRs, fosters innovation, and balances the interests of inventors, industries, and the public [101]. This involves multiple entities, processes, and frameworks operating at national and international levels.

Key administrative bodies

- **National patent offices**
  These include entities like the United States Patent and Trademark Office (USPTO), the European Patent Office (EPO), and the Indian Patent Office. They are responsible for examining, granting, and maintaining patents within their jurisdiction.
- **World Intellectual Property Organization (WIPO)**
  WIPO facilitates international cooperation in patents through systems like the PCT, enabling streamlined filing in multiple countries.
- **Regional patent offices**
  Examples include the EPO and the African Regional Intellectual Property Organization (ARIPO), which handle patent applications across member states.

Core responsibilities of patent administration

- **Patent application processing**
  Reviewing applications to ensure compliance with legal and technical standards.

- **Examination and granting**
  Conducting substantive examinations to determine novelty, non-obviousness, and industrial applicability.
- **Publication**
  Ensuring timely publication of patent applications to inform the public and promote transparency.
- **Opposition and appeals**
  Providing mechanisms for third parties to challenge patents before or after the grant.

Policy formulation and enforcement

- **Legislative updates**
  Administrators work with governments to update patent laws to reflect technological advancements and societal needs.
- **IP enforcement**
  Assisting in the resolution of patent disputes and combating infringement through legal frameworks.

International cooperation

- **Harmonization of patent laws**
  Initiatives like the TRIPS Agreement (Trade-Related Aspects of IPRs) promote uniform standards across countries.
- **PCT system**
  Allows inventors to seek patent protection in multiple countries with a single application.

Public awareness and support

- **Training and outreach**
  Educating inventors, businesses, and the public about patent processes and benefits.
- **Support for SMEs and startups**
  Offering reduced fees and advisory services to encourage innovation from smaller entities.

Technological integration

- **Digital systems for patent filing and examination**
  Online portals like ePCT and USPTO's EFS-Web streamline the patent process.
- **AI in patent searches**
  Use of artificial intelligence to enhance prior art searches and improve examination efficiency.

Challenges in patent administration

- **Backlogs**
  High application volumes can lead to delays in examination and granting.
- **Global disparities**
  Variations in patent laws and enforcement mechanisms across countries can complicate international filings.
- **Ethical and technological issues**
  Emerging fields like biotechnology and AI require updates to existing patent frameworks.

Efficient administration of the patent system is vital for fostering innovation, protecting inventors' rights, and contributing to economic growth. By continuously adapting to technological advancements and global challenges, patent offices and organizations ensure the relevance and effectiveness of the system in a dynamic world.

5.10 IPR OF BIOLOGICAL SYSTEMS

The protection of biological systems through IPRs has become increasingly important with advancements in biotechnology, genetics, and agricultural sciences. Biological systems include living organisms, genetic materials, and processes derived from them [102]. IPR ensures that innovators in these fields can secure their discoveries, fostering research and innovation while raising ethical and regulatory considerations.

Key forms of IPR for biological systems

- **Patents**
  - Protect inventions involving biological materials, processes, and genetically modified organisms (GMOs).
  - Examples include genetically engineered crops (e.g., Btcotton) and biopharmaceuticals like insulin-producing bacteria.
- **Plant variety protection (PVP)**
  - Recognizes breeders' rights over new plant varieties that meet criteria of novelty, distinctiveness, uniformity, and stability.
  - Encourages innovation in agriculture while safeguarding farmers' rights.
- **Trade secrets**
  - Protect confidential processes or formulations, such as proprietary methods for extracting bioactive compounds.
- **Geographical indications (GI)**
  - Protects traditional biological products associated with specific regions, such as Basmati rice or Darjeeling tea.

Applications in biological systems

- **Agricultural biotechnology**
  - Genetically modified crops, pest-resistant plants, and improved seed varieties.
  - Example: Monsanto's Roundup Ready crops, protected through patents.
- **Pharmaceuticals and biologics**
  - Use of biological systems to produce medicines, such as vaccines or monoclonal antibodies.
  - Example: Patented processes for producing recombinant human insulin.
- **Genomics and synthetic biology:**
  - Patents on specific genes, DNA sequences, or synthetic organisms with industrial or medical applications.
  - Example: CRISPR-Cas9 gene-editing technology.
- **Environmental biotechnology:**
  - Use of biological systems for waste management, pollution control, and renewable energy production.
  - Example: Patented microbes for bioremediation.

Challenges in protecting biological systems

- **Ethical concerns**
  - Ownership of genetic resources and life forms raises moral and societal questions.
  - Example: Opposition to patents on human genes or indigenous plant varieties.
- **Biopiracy**
  - The exploitation of biological resources or traditional knowledge without proper acknowledgement or benefit-sharing.
  - Example: Controversies surrounding neem, turmeric, and basmati rice patents.
- **Access and benefit-sharing (ABS)**
  - Compliance with the Nagoya Protocol ensures fair distribution of benefits derived from genetic resources.
- **Global disparities**
  - Developing countries often lack the resources to enforce or benefit from IPR protections in biological systems.

Future directions

- **International collaboration**
  - Harmonizing IPR laws to address cross-border challenges in biological innovations.

- **Balancing innovation and access**
  - Policies that promote research while ensuring equitable access to biological resources and benefits.
- **Digital sequence information (DSI)**
  - Addressing the IPR implications of genetic data stored in digital formats.

IPR in biological systems is a dynamic field that bridges innovation, ethics, and global cooperation. Its effective implementation can drive advancements in health, agriculture, and environmental sustainability while ensuring fair practices and equitable benefits.

5.11 IPR OF COMPUTER SOFTWARE

The rapid evolution of the software industry has underscored the need for robust IP protection to safeguard innovation, ensure fair competition, and foster technological advancement [103]. Software IPR encompasses legal mechanisms that protect the development, use, and distribution of computer programs and related intellectual assets.

Forms of IPR for computer software

- **Copyrights**
  - The most common form of protection for software.
  - Covers the source code, object code, and user interfaces.
  - Example: A video game's code and visual design are protected under copyright laws.
- **Patents**
  - Protect software-related inventions that demonstrate novelty, non-obviousness, and industrial applicability.
  - Examples include algorithms, data processing techniques, and innovative software functionalities.
  - Example: Amazon's patented "1-Click" online shopping process.
- **Trade secrets**
  - Protects confidential software elements, such as proprietary algorithms or unique development processes.
  - Example: Google's search engine algorithm.
- **Trademarks**
  - Protects the branding elements of software, such as logos, names, or slogans.
  - Example: The Microsoft Windows logo.
- **Licensing agreements**
  - Define the terms of use for software, ensuring the IPRs of the developer are upheld.
  - Examples include end-user license agreements (EULAs) and open-source licenses.

Key applications of software IPR

- **Commercial software**
  - Proprietary applications like Microsoft Office or Adobe Photoshop, protected through copyrights and patents.
- **Open-source software**
  - Software like Linux or Apache is distributed with open-source licenses that specify usage and modification rights.
- **Embedded software**
  - Software integrated into devices like smart appliances, vehicles, and medical equipment.
  - Example: Tesla's autopilot software protected through trade secrets and patents.
- **Artificial intelligence and machine learning**
  - Protection of AI algorithms, training datasets, and generated models.
  - Example: Patents on AI-driven voice recognition systems such as Amazon Alexa.

Challenges in protecting software IPR

- **Global variations in laws**
  - Differences in software patentability across jurisdictions.
  - Example: The United States allows software patents more broadly than the EU.
- **Piracy and unauthorized use**
  - The high prevalence of software piracy undermines IP protections.
  - Example: Illegally downloaded versions of paid software.
- **Difficulty in demonstrating novelty**
  - Establishing the originality of algorithms or functionalities for patent protection.
- **Overlapping protections**
  - Determining the most suitable IPR mechanism for a software product can be complex.
- **Rapid technological change**
  - The fast pace of software innovation can render IP protections obsolete quickly.

Best practices for software IPR

- **Comprehensive documentation**
  - Maintain detailed records of the software development process to establish ownership.
- **IPR strategy alignment**
  - Choose the appropriate form of IPR based on the software's purpose and market.

- **Use of licensing models**
  - Tailor licensing agreements to balance accessibility with IP protection.
- **Periodic Audits**
  - Conduct regular reviews to ensure compliance with IP laws and address potential infringements.
- **Leveraging international treaties**
  - Use agreements like the Berne Convention and TRIPS for cross-border IP protection.

Future trends in software IPR

- **AI-Generated software protection**
  - Addressing ownership issues for software created by artificial intelligence.
- **Blockchain in IPR**
  - Using blockchain for secure and transparent software licensing and IP management.
- **Global harmonization**
  - Efforts to unify software IP laws to simplify international protections.

Software IPR is pivotal in fostering innovation while protecting developers' rights and ensuring fair market practices. As the digital landscape evolves, balancing protection with accessibility will remain a cornerstone of effective IPR strategies in the software domain.

5.12 CASE STUDIES IN IPR

Case studies provide insight into the practical applications of IP laws, showcasing how they influence innovation, business strategies, and legal challenges. Below are notable examples of IPR in action.

5.12.1 Case study 1: Apple vs. Samsung—design and utility patents

The legal battle between Apple Inc. and Samsung Electronics Co. over design and utility patents is one of the most prominent IPR disputes in the technology industry [107]. This case highlights the significance of protecting innovation while demonstrating the complexities of enforcing IPR in a global market.

Background of the Case

- **Year initiated:** 2011
- **Parties involved**

  - **Apple Inc.**: Claimed that Samsung infringed on its patents related to the design and functionality of its iPhone.
  - **Samsung electronics:** Denied infringement and countersued Apple, alleging violations of its patents.
- **Patents in dispute**
  - **Design patents:** Covering the look and feel of Apple's devices, such as the rounded corners, bezels, and the grid layout of icons.
  - **Utility patents:** Related to functional features like pinch-to-zoom and bounce-back effects when scrolling.

Key issues in the case

- **Design patents**
  - Apple claimed Samsung's Galaxy devices mimicked the appearance of the iPhone, leading to consumer confusion.
  - Samsung argued that these design elements were functional and not unique to Apple.
- **Utility patents**
  - Apple accused Samsung of copying innovative features that enhance user interaction.
  - Samsung contested the validity of Apple's patents and argued that similar functionalities predated Apple's claims.
- **Damages and market impact**
  - Apple sought significant damages for the alleged infringement, claiming loss of market share and brand value.
  - Samsung contended that any similarities were superficial and did not warrant such high compensation.

Legal proceedings and outcomes

- **Initial verdict** (2012)
  - A US jury awarded Apple $1.05 billion in damages, ruling that Samsung had infringed on Apple's design and utility patents.
- **Appeals and adjustments:**
  - Over the years, the case went through multiple appeals, with damages being recalculated.
  - In 2018, Samsung agreed to pay Apple $539 million as a final settlement.
- **International litigation:**
  - Similar lawsuits were filed in other countries, with varying outcomes.
  - Some courts sided with Apple, while others ruled in favour of Samsung or dismissed the cases.

Key takeaways from the case

- **Importance of design patents**
  - The case underscored the value of protecting design elements as IP, especially in competitive industries like consumer electronics.
- **Challenges in utility patents**
  - Establishing the novelty and non-obviousness of functional features can be contentious and subject to interpretation.
- **Global IPR enforcement**
  - The case highlighted the difficulties of enforcing IPR across jurisdictions with differing legal frameworks.
- **Market and brand impact**
  - Beyond legal outcomes, the dispute influenced public perception and shaped the strategies of other tech companies regarding IPR.

Impact on the industry

1. **Increased focus on IPR**
   - Companies became more vigilant in protecting their IP, leading to an increase in patent filings.
2. **Innovation and competition**
   - The case demonstrated the balance between encouraging innovation and fostering healthy competition.
3. **Legal precedents**
   - The rulings set benchmarks for design and utility patent cases, influencing future IPR disputes.

The Apple vs. Samsung case serves as a landmark example of the complexities of IPR in the tech industry. It emphasizes the need for companies to protect their innovations while navigating the challenges of global enforcement. This case also highlights the evolving interplay between creativity, functionality, and legal frameworks in shaping the competitive landscape of modern technology.

5.12.2 Case study 2: Coca-Cola—trade secrets

The Coca-Cola Company's formula for its iconic beverage is one of the most famous examples of a trade secret [105]. This case demonstrates how trade secrets can provide a competitive edge while highlighting the importance of safeguarding proprietary information in a highly competitive market.

Background of the Case

- **The trade secret**
  The exact formula for Coca-Cola, known as "Merchandise 7X," has been a closely guarded secret since the drink's invention in 1886 by Dr. John Stith Pemberton.

- **Significance**
  The formula is central to Coca-Cola's brand identity and competitive advantage in the global beverage market.

Key issues in trade secret protection

- **Maintaining confidentiality**
  - The formula is reportedly known to only a few employees at any given time.
  - The company employs strict protocols to ensure secrecy, including storing the written formula in a high-security vault in Atlanta, Georgia.
- **Risk of misappropriation**
  - Trade secrets are vulnerable to theft, leaks, or reverse engineering.
  - Coca-Cola has faced attempts to uncover its formula, necessitating robust legal and security measures.
- **Legal protection**
  - Unlike patents, trade secrets are not disclosed publicly and do not have an expiration date, provided they remain confidential.
  - Coca-Cola relies on the **Uniform Trade Secrets Act (UTSA)** and other international laws to protect its IP.

Case of misappropriation attempt

In **2006**, three individuals, including a Coca-Cola employee, attempted to sell confidential company information, including details about new products, to PepsiCo.

- **Incident details**
  - The conspirators approached PepsiCo, offering the stolen information.
  - PepsiCo informed Coca-Cola and the authorities, showcasing ethical business practices.
- **Outcome**
  - The culprits were arrested and convicted.
  - The incident reinforced the importance of internal controls and ethical corporate behaviour.

Key Lessons from Coca-Cola's approach

- **Long-term competitive advantage**
  - The Coca-Cola formula demonstrates how trade secrets can sustain a brand's dominance for over a century without requiring periodic renewal, as patents do.
- **Organizational discipline**
  - Strict internal protocols and limited access are critical to maintaining the confidentiality of trade secrets.

- **Legal and ethical frameworks**
  - Coca-Cola's reliance on trade secret laws and PepsiCo's ethical handling of the misappropriation attempt highlights the interplay between legal measures and corporate ethics.
- **Innovation beyond the formula**
  - Coca-Cola's success is not solely based on the formula but also on its marketing, distribution, and customer engagement strategies.

The Coca-Cola case exemplifies the strategic value of trade secrets in preserving a competitive edge. By prioritizing confidentiality and leveraging robust security measures, Coca-Cola has maintained its market leadership and brand legacy. This case also underscores the ethical responsibilities of corporations in respecting competitors' IP and the critical role of legal frameworks in protecting trade secrets.

5.12.3 Case study 3: Novartis vs. India—pharmaceutical patents

The landmark legal battle between Swiss pharmaceutical company Novartis AG and the Government of India over the cancer drug *Glivec* (also known as *Gleevec*) highlighted key issues surrounding pharmaceutical patents, access to medicine, and IPRs in developing countries [106].

Background of the Case

- **The drug**
  Glivec is a life-saving drug used to treat chronic myeloid leukaemia (CML) and other cancers. Developed by Novartis, it was recognized for its effectiveness and innovation in targeting cancer cells.
- **The patent application**
  Novartis applied for a patent in India in 1998 for an improved version of the drug, which involved the beta-crystalline form of imatinib mesylate, the active ingredient in *Glivec.*
- **Indian patent law**
  Section 3(d) of the Indian Patents Act prohibits patents for incremental innovations unless they show a significant improvement in therapeutic efficacy.

Key issues in the case

- **Evergreening**
  - Critics argued that Novartis was attempting to extend its monopoly by patenting a slightly modified version of the drug, a practice known as "evergreening."
  - India's patent law aimed to prevent such practices to ensure affordable access to medicines.

- **Access to affordable medicines**
  - *Glivec* was priced at approximately $2,600 per month, making it unaffordable for most Indian patients.
  - Generic versions of the drug, sold by Indian companies, cost about $200 per month, significantly increasing accessibility.
- **Global vs. local perspectives**
  - Novartis contended that denying the patent discouraged innovation and harmed the pharmaceutical industry.
  - India argued that its laws were designed to balance innovation incentives with public health needs.

Legal proceedings

- **Patent office rejection (2006)**
  - India's Patent Office rejected Novartis's application, citing Section 3(d) and the lack of significant therapeutic efficacy in the new version.
- **Supreme court ruling (2013)**
  - The Supreme Court of India upheld the rejection, stating that Novartis's improved version did not meet the threshold of enhanced therapeutic efficacy required under Section 3(d).

Impact of the case

- **On pharmaceutical patents**
  - The ruling reinforced India's stance against evergreening, setting a precedent for rejecting similar patent applications.
- **On access to medicines**
  - The decision allowed Indian companies to continue producing affordable generic versions of *Glivec*, benefiting thousands of cancer patients.
- **On global pharma**
  - The case highlighted tensions between multinational pharmaceutical companies and countries prioritizing public health over IP protections.
- **On innovation**
  - Critics argued that the ruling might deter companies from investing in drug research for markets with stringent patent laws.

Key lessons

- **Balancing IP and public health**
  - The case underscores the need to balance innovation incentives with the broader goal of ensuring access to essential medicines.
- **Significance of national laws**
  - India's interpretation of Section 3(d) demonstrated the role of national laws in shaping the global IP landscape.

- **The role of generic medicines**
  - Affordable generics play a crucial role in addressing healthcare disparities in developing countries.
- **Corporate responsibility**
  - Pharmaceutical companies must consider pricing strategies that balance profitability with accessibility in low-income markets.

The *Novartis vs. India case* remains a milestone in the debate over pharmaceutical patents and public health. It reaffirmed the principle that IPRs must not come at the cost of affordable healthcare. By prioritizing access to medicines over patent protection, India's decision resonated globally, influencing policies in other countries with similar healthcare challenges.

5.12.4 Case study 4: Harry Potter series—copyright protection

The *Harry Potter* series, authored by JK Rowling, stands as one of the most celebrated and commercially successful literary works of all time. Beyond its literary acclaim, the series has also become a focal point for numerous copyright issues, highlighting the complexities of IP protection in creative works [107].

Background of the case

- **The work**
  The *Harry Potter* series includes seven books published between 1997 and 2007. The series has since expanded into movies, merchandise, theme parks, and spin-offs, creating a multibillion-dollar franchise.
- **Copyright protection**
  As the creator, JK Rowling holds the copyright to the text and characters, while publishers like Bloomsbury (United Kingdom) and Scholastic (United States) hold rights to specific editions. Warner Bros. holds copyright for film adaptations and related merchandise.

Key copyright issues

- **Unauthorized derivative works**
  - Numerous unauthorized sequels, prequels, and fan-fiction works have been created, leading to legal action to protect the integrity of the original works.
  - For example, *The Harry Potter Lexicon* by Steven Vander Ark was deemed an infringement due to its extensive use of Rowling's text and characters.
- **Piracy**
  - Digital piracy has been a persistent issue, with illegal copies of books and movies distributed online.

    - Rowling and her publishers have actively worked to combat piracy, leveraging technology and legal tools.
- **Trademark infringement**
    - The *Harry Potter* name and related terms have been trademarked. Unauthorized use of these trademarks for merchandise or events has resulted in legal action.
- **Fan works vs. copyright enforcement**
    - While Rowling has been supportive of fan fiction and fan art under certain conditions, commercial exploitation of such works without permission has been prohibited.

Legal proceedings

- **The harry potter lexicon case (2008)**
    - JK Rowling sued RDR Books for publishing *The Harry Potter Lexicon*, arguing that it infringed on her copyright by extensively copying her text.
    - The court ruled in Rowling's favour, stating that while reference works are permissible, they cannot replicate or replace the original material.
- **Trademark disputes**
    - Warner Bros. has taken legal action against individuals and companies using *Harry Potter* trademarks for unauthorized purposes, such as themed events or products.

Impact of the case

- **Strengthened copyright protection**
    - The legal actions reinforced the importance of copyright in protecting creative works and maintaining their value.
- **Balance with fan engagement**
    - Rowling's selective approach to fan-created content highlights the need to balance legal enforcement with fostering a loyal fan base.
- **Global awareness of copyright**
    - The widespread popularity of *Harry Potter* brought attention to copyright laws and their role in protecting IP.
- **Piracy countermeasures**
    - The case emphasized the challenges of combating piracy in the digital age, spurring advancements in anti-piracy technology.

Key lessons

- **Protecting creative integrity**
    - Copyright ensures that creators maintain control over how their works are used and prevents unauthorized exploitation.

- **The role of fair use**
  - While derivative works may fall under fair use, they must not replicate or replace the original material.
- **Trademark and branding**
  - Protecting a franchise extends beyond copyright to include trademarks, ensuring comprehensive IP protection.
- **Fan interaction**
  - Creators and companies must find a balance between protecting IP and fostering community engagement, as fan works can enhance a franchise's longevity.

The *Harry Potter* copyright cases illustrate the complexities of IP protection in the entertainment industry. By defending her work, JK Rowling has not only safeguarded her creative rights but also set a precedent for protecting literary and cultural properties in a globalized, digital world. This case highlights the ongoing need for vigilance in enforcing copyright while navigating the fine line between legal protection and fan engagement.

5.12.5 Case study 5: Darjeeling Tea—geographical indications (GI)

Darjeeling Tea, known for its unique flavour, aroma, and quality, is one of the most renowned examples of GI in the world [108]. The GI status granted to Darjeeling Tea protects its authenticity, prevents misuse of its name, and ensures economic benefits for producers in the region.

Background of the case

- **The product**
  - Darjeeling Tea is grown exclusively in the Darjeeling district of West Bengal, India.
  - Its distinct flavour is attributed to the region's specific climatic conditions, altitude, and soil quality.
- **Geographical indication status**
  - In 2004, Darjeeling Tea became the first product in India to receive a GI tag under the GIs of Goods (Registration and Protection) Act, 1999.
- **Global importance**
  - Darjeeling Tea is often referred to as the "Champagne of Teas" due to its global reputation and premium status.

Challenges before GI recognition

- **Misuse of the name**
  - The term "Darjeeling Tea" was widely used for teas grown outside the Darjeeling region, misleading consumers and affecting the authenticity of the product.

- **Economic loss**
  - Unauthorized use of the name diluted the market, causing financial losses for legitimate Darjeeling Tea producers.
- **Lack of consumer awareness**
  - Many consumers were unaware of the specific characteristics that make Darjeeling Tea unique, further complicating its market positioning.

The role of GI protection

- **Authenticity assurance**
  - The GI tag certifies that only tea produced in the 87 designated gardens in Darjeeling, meeting specific quality standards, can be sold as "Darjeeling Tea."
- **Consumer trust**
  - GI status helps consumers identify genuine Darjeeling Tea, ensuring they receive a product with the promised quality and origin.
- **Legal safeguards**
  - Producers gained the legal right to prevent unauthorized use of the Darjeeling Tea name, both domestically and internationally.

Notable legal disputes

- **European union (EU) recognition**
  - In 2011, the EU granted Protected Geographical Indication (PGI) status to Darjeeling Tea, ensuring its protection within EU markets.
- **Trademark conflicts**
  - The Tea Board of India, which oversees the GI certification, has actively pursued cases against companies falsely labelling their products as Darjeeling Tea, achieving several legal victories.

Economic and social impact

- **Economic benefits**
  - GI recognition has allowed Darjeeling Tea producers to command premium prices, boosting the local economy.
- **Cultural preservation:**
  - Protecting Darjeeling Tea also safeguards the region's cultural heritage and traditional tea cultivation practices.
- **Sustainable practices**
  - The GI framework encourages sustainable agricultural practices, ensuring the long-term viability of Darjeeling Tea production.

Key lessons

- **Importance of geographical indications**
  - GI tags provide a robust framework for protecting the identity of region-specific products, ensuring authenticity and quality.

- **Economic empowerment**
  - GI recognition can significantly enhance the economic prospects of producers by preventing market dilution and enabling premium pricing.
- **Global branding**
  - International recognition of GI products like Darjeeling Tea highlights the role of IP in promoting cultural and regional identities globally.
- **Consumer Awareness**
  - Educating consumers about the value and uniqueness of GI-tagged products is essential for their success in the market.

Darjeeling Tea's GI recognition underscores the importance of protecting GIs in a globalized economy. It demonstrates how legal safeguards can preserve cultural heritage, ensure product authenticity, and promote economic development. The success of Darjeeling Tea as a GI-tagged product serves as a model for other region-specific goods, emphasizing the potential of IP to empower local communities while protecting their unique identities.

5.12.6 Case study 6: Google vs. Oracle—software copyright

The Google vs. Oracle case is a landmark legal battle in IP law, specifically focusing on software copyright. It centred around the use of Java Application Programming Interface (API) by Google in its Android operating system, raising critical questions about the boundaries of copyright protection for software [109].

Background of the case

- **The dispute**
  - Oracle acquired Sun Microsystems in 2010, obtaining rights to Java.
  - Google used parts of Java APIs to develop Android, particularly 37 API packages, claiming it was "fair use."
- **The conflict**
  - Oracle alleged copyright infringement, seeking billions in damages, asserting that Google's use of Java APIs without a license was unlawful.
  - Google countered that APIs are functional and therefore not eligible for copyright protection.

Key legal questions

- **Are APIs copyrightable?**

    - Oracle argued that Java APIs are creative works and thus subject to copyright protection.
    - Google maintained that APIs are functional and essential for interoperability, making them ineligible for copyright.
- **Was google's use "fair use"?**
    - Google contended that its use of the APIs constituted "fair use" because it only used what was necessary to develop a transformative product (Android).

Legal proceedings and outcomes

- **District court ruling** (2012)
    - The court ruled in favour of Google, stating that APIs are not copyrightable.
- **Appeals court ruling** (2014)
    - The Federal Circuit reversed the decision, holding that APIs are copyrightable and remanded the case to determine fair use.
- **Supreme court decision** (2021)
    - The U.S. Supreme Court ruled 6-2 in favour of Google, stating that its use of Java APIs was fair use.
    - The court emphasized the transformative nature of Android and the limited scope of the copied code.

Significance of the case

- **Impact on software development**
    - The decision affirmed that certain uses of APIs could be considered fair use, fostering innovation and interoperability in software development.
- **Clarification on copyright law**
    - The case highlighted the complexity of applying copyright law to functional works like software, setting a precedent for future disputes.
- **Industry implications**
    - A ruling in Oracle's favour could have led to stricter restrictions on API usage, potentially stifling innovation and increasing costs for developers.

Lessons from the case

- **Fair use doctrine**
    - The ruling reinforced the importance of the fair use doctrine in promoting innovation, especially in transformative uses of copyrighted material.
- **Balance between protection and innovation**
    - The case demonstrated the need to balance IP protection with the freedom to build upon existing technologies for societal benefit.

- **Evolving nature of copyright law**
  - It highlighted the challenges of applying traditional copyright principles to rapidly evolving fields like software and technology.

The Google vs. Oracle case is a milestone in software copyright law, shaping how APIs are treated under IP law. It underscores the delicate balance between protecting creators' rights and fostering innovation. By ruling in favour of fair use, the Supreme Court provided clarity on the legal boundaries of software development, ensuring that innovation remains a driving force in the tech industry.

5.13 SUMMARY

This chapter delves into the complexities of patent rights and their role in IP management, starting with the **scope of patent rights** and their importance in safeguarding innovations. It explores **licensing and technology transfer,** discussing the types of licensing agreements, their benefits, and the mechanisms of transferring technology between entities. A detailed section on **patent licensing in technology transfer** emphasizes its significance in commercializing innovations.

The chapter also examines the **challenges in licensing and technology transfer**, including legal, financial, and logistical hurdles, and highlights the critical role of **TTOs** in facilitating smooth IP transactions. The importance of **patent information and databases** for research, analysis, and strategy development is also covered.

Emerging trends in **new developments in IPR** and their implications on **biological systems** and **computer software** are explored, reflecting the evolving nature of IP in cutting-edge fields. The chapter concludes with an array of **case studies** illustrating real-world scenarios, including landmark cases like **Apple vs. Samsung** on design and utility patents, **Coca-Cola** and its trade secrets, **Novartis vs. India** on pharmaceutical patents, **Harry Potter series** for copyright protection, **Darjeeling Tea** under GIs, and **Google vs. Oracle** on software copyright.

By combining theoretical insights with practical examples, this chapter provides a comprehensive understanding of patent rights, licensing, technology transfer, and the broader landscape of IP management.

REVIEW QUESTIONS

1. What is the scope of patent rights, and how do they protect intellectual property?
2. Explain the significance of patent rights in fostering innovation.
3. Define licensing and discuss its role in intellectual property management.

4. What are the different types of licensing, and what are their key benefits?
5. Explain the concept of technology transfer and its importance in innovation dissemination.
6. How does patent licensing facilitate technology transfer?
7. What challenges are commonly faced during patent licensing and technology transfer?
8. What is the role of Technology Transfer Offices (TTOs) in IP management?
9. How do TTOs contribute to overcoming challenges in technology transfer?
10. Why are patent information and databases essential for research and IP strategy?
11. What types of data can be accessed through patent databases?
12. What are the recent developments in intellectual property rights?
13. How do new IPR trends impact industries like biotechnology and software?
14. Discuss the challenges and opportunities in protecting intellectual property for biological systems.
15. How is intellectual property managed in the context of computer software?
16. What are the primary challenges in enforcing software copyrights?
17. Summarize the key issues and outcomes of the Apple vs. Samsung case on design and utility patents.
18. What lessons can be learned from the Coca-Cola case on trade secrets?
19. Discuss the implications of the Novartis vs. India case on pharmaceutical patents.
20. How does the Harry Potter series illustrate copyright protection in creative industries?
21. What makes Darjeeling Tea a notable example of geographical indications?
22. How do licensing and technology transfer contribute to the commercialization of innovations?
23. Why is it important to address challenges in patent licensing and technology transfer?
24. Discuss the role of case studies in understanding the practical implications of IPR.

Bibliography

1. Smith, B., & McGannon, K. R. (2018). Developing rigor in qualitative research: Problems and opportunities within sport and exercise psychology. *International review of sport and exercise psychology*, *11*(1), 101–121, https://doi.org/10.1080/1750984X.2017.1317357
2. Favero, N., & Bullock, J. B. (2015). How (not) to solve the problem: An evaluation of scholarly responses to common source bias. *Journal of Public Administration Research and Theory*, *25*(1), 285–308, https://doi.org/10.1093/jopart/muu020
3. Norman, G. R. (1988). Problem-solving skills, solving problems and problem-based learning. *Medical education*, 22(4), 279–286, https://doi.org/10.1111/j.1365-2923.1988.tb00754.x
4. Fox, J. H. (1958). Criteria of good research. *The Phi Delta Kappan*, *39*(6), 284–286.
5. Yadav, D. (2022). Criteria for good qualitative research: A comprehensive review. *The Asia-Pacific Education Researcher*, *31*(6), 679–689, https://doi.org/10.1007/s40299-021-00619-0
6. Mohajan, H. K. (2017). Two criteria for good measurements in research: Validity and reliability. *Annals of Spiru Haret University. Economic Series*, *17*(4), 59–82, https://doi.org/10.26458/1746
7. Denscombe, M. (2017). *EBOOK: The good research guide: For small-scale social research projects*. 7th Edition; ISBN: 9780335249831; Education (UK).
8. Nathan, P. E., & Lansky, D. (1978). Common methodological problems in research on the addictions. *Journal of Consulting and Clinical Psychology*, *46*(4), 713, https://doi.org/10.1037/0022-006X.46.4.713
9. Haddaway, N. R., Bethel, A., Dicks, L. V., Koricheva, J., Macura, B., Petrokofsky, G., ... & Stewart, G. B. (2020). Eight problems with literature reviews and how to fix them. *Nature Ecology & Evolution*, *4*(12), 1582–1589, https://doi.org/10.1038/s41559-020-01295-x
10. Theofanidis, D., & Fountouki, A. (2018). Limitations and delimitations in the research process. *Perioperative Nursing-Quarterly scientific, online official journal of GORNA*, *7*(3 September-December 2018), 155–163, https://doi.org/10.5281/zenodo.2552021
11. Akanle, O., Ademuson, A. O., & Shittu, O. S. (2020). Scope and limitation of study in social research. *Contemporary issues in social research*, 105–114, Ibadan University Press, University of Ibadan, Ibadan, Nigeria.

12. Deevy, M., & Beals, M. (2013). The scope of the problem. *Financial Fraud Research Center*, Stanford Center on Longevity, CA, USA.
13. Doody, O., & Bailey, M. E. (2016). Setting a research question, aim and objective. *Nurse researcher*, *23*(4), https://doi.org/10.7748/nr.23.4.19.s5
14. Huan, S. (2004). Objectives and methods of research-oriented environmental education. *Chinese Education & Society*, *37*(4), 57–63, https://doi.org/10.1080/10611932.2004.11031647
15. Bairagi, V., & Munot, M. V. (Eds.). (2019). *Research methodology: A practical and scientific approach*. Chapman and Hall/CRC Press, New York. https://doi.org/10.1201/9781351013277
16. Bang, T. C. (2024). Ensuring Credibility and Trustworthiness in Qualitative Inquiries. In *Applied Linguistics and Language Education Research Methods: Fundamentals and Innovations* (pp. 70–85). IGI Global, PA, USA.
17. Colnerud, G. (2006). Teacher ethics as a research problem: Syntheses achieved and new issues. *Teachers and Teaching: theory and practice*, *12*(3), 365–385, https://doi.org/10.1080/13450600500467704
18. Dworkin, S. F. (1992). Approach to the problem. *Journal of Craniomandibular Disorders*, *6* (4), 302–306.
19. Gallagher, M. (2009). Data collection and analysis. In *Researching with Children and Young People: Research Design, Methods and Analysis* (pp. 65–153), SAGE Publications Ltd, https://doi.org/10.4135/9781446268315
20. Mazhar, S. A., Anjum, R., Anwar, A. I., & Khan, A. A. (2021). Methods of data collection: A fundamental tool of research. *Journal of Integrated Community Health*, *10*(1), 6–10, https://doi.org/10.24321/2319.9113.202101
21. Ellis, T. J., & Levy, Y. (2008). Framework of problem-based research: A guide for novice researchers on the development of a research-worthy problem. *Informing Science*, *11*, 17, https://doi.org/10.28945/438
22. Loewenstein, R. M. (1951). The problem of interpretation. *The Psychoanalytic Quarterly*, *20*(1), 1–14, https://doi.org/10.1080/21674086.1951.11925827
23. Birmingham, P., & Wilkinson, D. (2003). *Using research instruments: A guide for researchers*. Routledge, London, https://doi.org/10.4324/9780203422991
24. Fink, A. (2019). *Conducting research literature reviews: From the internet to paper*. Sage publications, Los Angeles, USA.
25. Snyder, H. (2019). Literature review as a research methodology: An overview and guidelines. *Journal of business research*, *104*, 333–339, https://doi.org/10.1016/j.jbusres.2019.07.039
26. Barber, W. E., & Taylor, J. N. (1990). The importance of goals, objectives, and values in the fisheries management process and organization: a review. *North American Journal of Fisheries Management*, *10*(4), 365–373, https://doi.org/10.1577/1548-8675(1990)010%3C0365:TIOGOA%3E2.3.CO;2
27. Levy, Y., & Ellis, T. J. (2006). A systems approach to conduct an effective literature review in support of information systems research. *Informing Science*, *9*, https://doi.org/10.28945/479
28. Booth, A., James, M. S., Clowes, M., & Sutton, A. (2021). *Systematic approaches to a successful literature review*, SAGE Publications Ltd, London, UK.
29. Templier, M., & Paré, G. (2015). A framework for guiding and evaluating literature reviews. *Communications of the Association for Information Systems*, *37*(1), 6, https://doi.org/10.17705/1CAIS.03706

30. Karunarathna, I., Gunasena, P., De Alvis, K., & Jayawardana, A. (2024). Structured reviews: Organizing, synthesizing, and analyzing scientific literature. *Retrieved from Research Gate.*
31. Rowlinson, M., Hassard, J., & Decker, S. (2014). Research strategies for organizational history: A dialogue between historical theory and organization theory. *Academy of Management Review, 39*(3), 250–274, https://doi.org/10.5465/amr.2012.0203
32. Wu, B., Wang, J., Cai, H., Shen, Y., Qu, B., & Fu, Y. (2024). Research on the Analysis Method of Production Safety Accidents Based on Accident Event Causal Association Diagram. *Mining, Metallurgy & Exploration, 41*(6), 1–14. https://doi.org/10.1007/s42461-024-01088-x
33. Keman, H., & Pennings, P. (2014). Comparative research methods. *Comparative politics, 3*, 47–59, Oxford University Press, Oxford UK.
34. Saiti, A., & Stefou, T. (2020). Hierarchical organizational structure and leadership. In *Oxford Research Encyclopedia of Education*. https://doi.org/10.1093/acrefore/9780190264093.013.709
35. Zina, O. (2021). *The essential guide to doing your research project*. Second Edition, SAGE Publications Pvt. Ltd, London, UK.
36. Paul, D. R. (n.d.) Plagiarism in Research Publications - An Overview. In *PCC* (p. 210).
37. Ridley, D. (2012). The literature review: A step-by-step guide for students, SAGE Publications Pvt. Ltd; London, UK.
38. Machi, L. A., & McEvoy, B. T. (2009). *The literature review: Six steps to success* Corwin (A SAGE Publication Company), California, USA.
39. Van, T. T. M. (2009). The relevance of literary analysis to teaching literature in the EFL Classroom. In *English teaching forum* (Vol. 47, No. 3, p. 2). US Department of State. Bureau of Educational and Cultural Affairs, Office of English Language Programs, SA-5, 2200 C Street NW 4th Floor, Washington, DC 20037.
40. Harris, D. (2019). *Literature review and research design: A guide to effective research practice*. Routledge, NY, USA. https://doi.org/10.4324/9780429285660-3
41. Velmurugan, V. S. (2024) Types and Definitions of Plagiarism: An Overview, *Global Research Journal of Social Sciences and Management* 2(1), 13–18. https://doi.org/10.55306/GRJSSM.2024.2102
42. Kumar, R., Pratap, B., & Pateria, R. K. (2022). An effect on academic integrity due to plagiarism: A review. *Library Progress (International), 42*(1), 206–219. https://doi.org/10.5958/2320-317x.2022.00019.8
43. Pecorari, D. (2013). *Teaching To Avoid Plagiarism: How To Promote Good Source Use: How to Promote Good Source Use*. McGraw-Hill Education, (UK), https://doi.org/10.1016/j.esp.2014.11.007
44. LaFollette, M. C. (1992). *Stealing into print: Fraud, plagiarism, and misconduct in scientific publishing*. Univ of California Press, https://doi.org/10.1525/9780520917804-013
45. Helgesson, G., & Eriksson, S. (2015). Plagiarism in research. *Medicine, Health Care and Philosophy, 18*, 91–101. https://doi.org/10.1007/s11019-014-9583-8
46. Brown, N., & Janssen, R. (2017). Preventing plagiarism and fostering academic integrity: a practical approach. *Journal of Perspectives in Applied Academic Practice, 5*(3), 102–109. https://doi.org/10.14297/jpaap.v5i3.245

47. Naik, R. R., Landge, M. B., & Mahender, C. N. (2015). A review on plagiarism detection tools. *International Journal of Computer Applications*, *125*(11), 16–22. https://doi.org/10.5120/ijca2015906113
48. Barrow, J. M., Brannan, G. D., & Khandhar, P. B. (2017). Research ethics, Stat Pearls Publishing, Treasure Island (FL), USA.
49. Fischer IV, B. A. (2006). A summary of important documents in the field of research ethics *Schizophrenia Bulletin*, *32*(1), 69–80. https://doi.org/10.1093/schbul/sbj005
50. Blake, G., & Bly, R. W. (1993). *The elements of technical writing*: Macmillan, NY, USA.
51. Lindsell-Roberts, S. (2023). *Technical writing for dummies*. John Wiley & Sons, USA.
52. Gerson, S. J., & Gerson, S. M. (2000). *Technical writing*. Prentice Hall, NJ, USA.
53. Dobrin, D. N. (1989). *Writing and Technique*. National Council of Teachers of English, Urbana, IL, USA.
54. Schneider, W. J., Lichtenberger, E. O., Mather, N., & Kaufman, N. L. (2018). *Essentials of assessment report writing*. John Wiley & Sons, USA.
55. Bowden, J. (2011). *Writing a report: how to prepare, write & present really effective reports*. Hachette UK.
56. Halwai, A., & Vaswani, V. (2023). Post-approval process: A challenge for ethics committees. *Perspectives in Clinical Research*, *14*(3), 139–145. https://doi.org/10.4103/picr.picr_214_22
57. Rao, P. S. (2017). The characteristics of effective writing skills in English language teaching. *Research Journal of English*, 2(2), 75–86.
58. Kallestinova, E. D. (2011). How to write your first research paper. *The Yale journal of biology and medicine*, *84*(3), 181. https://doi.org/10.1109/iedec.2013.6526784
59. Newsome, B. O. (2015). *An introduction to research, analysis, and writing: Practical skills for social science students*. Sage Publications, https://doi.org/10.4135/9781071909829.n11
60. Marsen, S. (2019). *Professional writing*. Bloomsbury Publishing, Bloomsbury, UK
61. Graf, C., Wager, E., Bowman, A., Fiack, S., Scott-Lichter, D., & Robinson, A. (2007). Best practice guidelines on publication ethics: a publisher's perspective. *International journal of clinical practice*, *61*, 1–26 https://doi.org/10.1111/j.1742-1241.2006.01230.x
62. Denney, A. S., & Tewksbury, R. (2013). How to write a literature review. *Journal of criminal justice education*, *24*(2), 218–234 https://doi.org/10.1080/10511253.2012.730617
63. Randaccio, M. (2013). Writing skills: theory and practice, https://doi.org/10.7748/cnp2013.09.12.7.12.s12
64. Ashbaugh, H., Johnstone, K. M., & Warfield, T. D. (2002). Outcome assessment of a writing-skill improvement initiative: results and methodological implications. *Issues in Accounting Education*, *17*(2), 123–148, https://doi.org/10.2308/iace.2002.17.2.123
65. Dhammi, I. K., & Haq, R. U. (2018). Standard format for writing a manuscript: A guide to authors. *Indian Journal of Orthopaedics*, *52*(4), 341, https://doi.org/10.4103/ortho.ijortho_323_18

66. Kafle, S., Yeung, P., & Huenerfauth, M. (2019, October). Evaluating the benefit of highlighting key words in captions for people who are deaf or hard of hearing. In *Proceedings of the 21st International ACM SIGACCESS Conference on Computers and Accessibility* (pp. 43–55), https://doi.org/10.1145/3308561.3353781
67. Dallimore, E. J., Hertenstein, J. H., & Platt, M. B. (2008). Using discussion pedagogy to enhance oral and written communication skills. *College Teaching, 56*(3), 163–172, https://doi.org/10.3200/ctch.56.3.163-172
68. Bisogno, M., Dumay, J., Manes Rossi, F., & Tartaglia Polcini, P. (2018). Identifying future directions for IC research in education: a literature review. *Journal of Intellectual Capital, 19*(1), 10–33, https://doi.org/10.1108/jic-10-2017-0133
69. Ondrusek, A. L. (2012). What the Research Reveals about Graduate Students' Writing Skills: A Literature Review. *Journal of Education for Library and Information Science, 53*(3), 176–188. http://www.jstor.org/stable/23249110
70. Savale, S. K., & Savale, V. K. (2016). Intellectual property rights (IPR). *World Journal of Pharmacy and Pharmaceutical Sciences, 5*(6), 2529–2559.
71. Kalanje, C. M. (2006). Role of intellectual property in innovation and new product development. *World Intellectual Property Organization.*
72. De Rassenfosse, G., Dernis, H., Guellec, D., Picci, L., & De La Potterie, B. V. P. (2013). The worldwide count of priority patents: A new indicator of inventive activity. *Research Policy, 42*(3), 720–737. https://doi.org/10.1016/j.respol.2012.11.002
73. Hall, B. H., & Harhoff, D. (2012). Recent research on the economics of patents. *Annual Review of Economics, 4*(1), 541–565. https://doi.org/10.1146/annurev-economics-080511-111008
74. Blackett, T. (2016). *Trademarks.* Macmillan Press Ltd., London, UK.
75. Varian, H. R. (2005). Copying and copyright. *Journal of economic perspectives, 19*(2), 121–138. https://doi.org/10.1257/0895330054048768
76. Brown, R. S. (1986). Design protection: An overview. *UCLA Law Review, 34*, 1341.
77. Steidman, J. C. (1962). Trade Secrets. *Ohio State Law Journal, 23*, 4.
78. Rangnekar, D. (2004). The socio-economics of geographical indications. *UNCTAD-ICTSD Project on IPRs and Sustainable Development, Issue Paper, 8*, 13–15.
79. Tripp, R., Louwaars, N., & Eaton, D. (2007). Plant variety protection in developing countries. A report from the field. *Food policy, 32*(3), 354–371. https://doi.org/10.1016/j.foodpol.2006.09.003
80. Chari, K. S., & Sharma, M. (2013, April). Integrated circuit layout design screening. In *2013 IEEE Conference on Information & Communication Technologies* (pp. 1305–1308). https://doi.org/10.1109/cict.2013.6558303
81. Ransley, D. L., & Gaffney, R. C. (1997). Upgrade your patenting process. *Research-Technology Management, 40*(3), 41–46. https://doi.org/10.1080/08956308.1997.11671129
82. Hunt, D., Nguyen, L., & Rodgers, M. (Eds.). (2012). *Patent searching: tools & techniques.* John Wiley & Sons, USA.
83. Sheldon, J. G. (2009). *How to write a patent application.* Practising Law Institute, New York, NY.

84. Wegner, H. C. (1992). Drafting the Patent Application. In *Patent Law in Biotechnology, Chemicals & Pharmaceuticals* (pp. 53–62). Palgrave Macmillan, London, UK. https://doi.org/10.1007/978-1-349-21958-2_7
85. Sampat, B. N., & Lemley, M. A. (2010). Examining patent examination, *Stanford Technology Law Review*, *2010*, 2, Stanford Public Law Working Paper No. 1485011.
86. Guellec, D., & de la Potterie, B. V. P. (2000). Applications, grants and the value of patent. *Economics letters*, *69*(1), 109–114, https://doi.org/10.1016/s0165-1765(00)00265-2
87. Jin, X., Spangler, S., Chen, Y., Cai, K., Ma, R., Zhang, L., ... & Han, J. (2011, December). Patent maintenance recommendation with patent information network model. In *2011 IEEE 11th International Conference on Data Mining* (pp. 280–289). https://doi.org/10.1109/icdm.2011.116
88. Burstein, M. J. (2014). Rethinking Standing in Patent Challenges. *The George Washington Law Review*, *83*, 498 & *Cardozo Legal Studies Research Paper No. 414*.
89. Holbrook, T. R. (2006). Possession in patent law, SMU Law Review, *59*, 123.
90. Paci, R., Sassu, A., & Usai, S. (1997). International patenting and national technological specialization. *Technovation*, *17*(1), 25–38. https://doi.org/10.1016/S0166-4972(96)00065-X
91. Jorgenson, L., & Fink, C. (2023). WIPO's Contributions to International Cooperation on Intellectual Property. *Journal of International Economic Law*, *26*(1), 30–34. https://doi.org/10.1093/jiel/jgac049
92. Naik, B. K. R., Jain, S. K., & Yadav, S. S. A. Study of Patent Cooperation Treaty (PCT) and the Impact of E-Technology.
93. Candelin-Palmqvist, H., Sandberg, B., & Mylly, U. M. (2012). Intellectual property rights in innovation management research: A review. *Technovation*, *32*(9-10), 502–512. https://doi.org/10.1016/j.technovation.2012.01.005
94. Merges, R. P., & Reynolds, G. H. (2000). The proper scope of the copyright and patent power. *Harvard Journal on Legislation*, *37*, 45.
95. Feldman, R., & Lemley, M. A. (2015). Do patent licensing demands mean innovation. *Iowa Law Review*, *101*, 137. https://doi.org/10.31235/osf.io/rfh6n
96. Lemley, M. A., & Feldman, R. (2016). Patent licensing, technology transfer, and innovation. *American Economic Review*, *106*(5), 188–192. https://doi.org/10.1257/aer.p20161092
97. Fitzgerald, J. D. (1992). Technology transfer issues in licensing pharmaceutical products. *R&D Management*, *22*(3), 199–208. https://doi.org/10.1111/j.1467-9310.1992.tb00810.x
98. Ustundag, A., Uğurlu, S., & Serdar Kilinc, M. (2011). Evaluating the performance of technology transfer offices. *Journal of Enterprise Information Management*, *24*(4), 322–337. https://doi.org/10.1108/17410391111148576
99. Singh, V., Chakraborty, K., & Vincent, L. (2016). Patent database: Their importance in prior art documentation and patent search. *Journal of Intellectual Property Rights*, *21*, 42–56. http://nopr.niscpr.res.in/handle/123456789/34016
100. Dolfsma, W. (2006). IPRs, technological development, and economic development. *Journal of Economic Issues*, *40*(2), 333–342. https://doi.org/10.1080/00213624.2006.11506911

101. Boehm, K., & Silberston, A. (1967). *British Patent System: Volume 1, Administration*, (Vol. *1*). CUP Archive.
102. Bera, R. K. (2015). Synthetic biology and intellectual property rights. *Biotechnology*. https://doi.org/10.5772/59906
103. Piera, F. (2003). IPR Protection of Computer Programs and Computer Software in the Global Market. *Currents: International Trade Law Journal*, *12*, 15.
104. Carani, C. V. (2012). Apple v Samsung-Design Patents Take Center Stage. *Landslide*, *5*, 25.
105. Halligan, R. M. (2009). Trade secrets v. patents: the new calculus. *Landslide*, 2, 10.
106. Gabble, R., & Kohler, J. C. (2014). To patent or not to patent? the case of Novartis' cancer drug Glivec in India. *Globalization and Health*, *10*, 1–6 https://doi.org/10.1186/1744-8603-10-3
107. Chatelain, M. (2012). Harry Potter and the prisoner of copyright law: Fan fiction, derivative works, and the fair use doctrine. *Tulane Journal of Technology and Intellectual Property*, *15*, 199.
108. Ravindran, S., & Mathew, A. (2009). The protection of geographical indication in india–case study on 'darjeeling tea'. *International Property Rights Index*, 58–64.
109. Gratz, J., & Lemley, M. A. (2017). Platforms and Interoperability in Oracle v Google. *Harvard Journal of Law & Technology*, *31*, 603.

Index

Pages in *italics* refer to figures and pages in **bold** refer to tables.